Concepts, Strategies and Models to Enhance Physics Teaching and Learning

Eilish McLoughlin · Paul van Kampen
Editors

Concepts, Strategies and Models to Enhance Physics Teaching and Learning

 Springer

Editors
Eilish McLoughlin
School of Physical Sciences & Centre
for the Advancement of STEM Teaching
and Learning
Dublin City University
Dublin, Ireland

Paul van Kampen
School of Physical Sciences & Centre
for the Advancement of STEM Teaching
and Learning
Dublin City University
Dublin, Ireland

ISBN 978-3-030-18136-9   ISBN 978-3-030-18137-6   (eBook)
https://doi.org/10.1007/978-3-030-18137-6

This Springer imprint is published by the registered company Springer Nature Switzerland AG
The registered company address is: Gewerbestrasse 11, 6330 Cham, Switzerland

Preface

The book presents papers selected under the leadership of GIREP vzw—the International Research Group on Physics Teaching, the organization promoting enhancement of the quality of physics teaching and learning at all educational levels and in all contexts. Through organization of annual conferences and seminars, active participation of researchers and practitioners in various GIREP Thematic Groups and wide cooperation with other international organizations involved in physics education, GIREP vzw facilitates the exchange of information and good practices in physics education, supports the improvement of the quality of pre-service and in-service professional development in physics teaching, promotes research in the field and facilitates the cooperation between stakeholders on both national and international levels.

This book is based on contributions presented at the GIREP-ICPE-EPEC 2017 conference, which was hosted in Dublin from 3 to 7 July 2017 by the Centre for the Advancement of STEM Teaching and Learning (CASTeL) at Dublin City University Ireland. This conference was organized by a collaboration between the International Research Group on Physics Teaching (GIREP); European Physical Society–Physics Education Division, and the Physics Education Commission of the International Union of Pure and Applied Physics (IUPAP). In total, 270 international researchers and practitioners from 48 countries participated in the conference and shared their knowledge and experiences under the theme of Bridging Research and Practice in Physics Teaching and Learning. A wide variety of topics and approaches, conducted in various contexts and settings, all adding a strong contribution to the field of physics education research were presented during the week-long conference. Examples include the design of curriculum and strategies to develop student competencies–including knowledge, skills, attitudes and values; inquiry workshop approaches in teacher education and pedagogical strategies adopted to engage and motivate students. Overall, a total of 271 contributions were presented—including 6 invited keynotes and 154 oral presentations, 40 symposia (each with four papers), 12 ICT demonstration/workshop sessions and 59 poster presentations. This book is built on 20 papers carefully selected in a rigorous double-blinded peer-review process, involving members of the editorial board and additional referees in order to guarantee the quality of the content of this contribution.

This collection of 20 chapters is presented in four parts, each with a focus on a particular aspect of research and practice in physics teaching and learning.

Part I: Development of Physics Teaching and Learning in School discusses a range of different models and strategies used to improve the teaching and learning of physics concepts in the classroom. Approaches include the introduction of the concept of voltage as an electric pressure difference across a resistor in an electric circuit, the use of graphical and visual analogies and models to introduce Einstein's General Theory of Relativity and a virtual sandbox which incorporates a combination of real-world experiments and real-time computer simulations to introduce the principles of granular flow physics. The Teaching Enquiry with Mysteries Incorporated (TEMI) approach designed for teaching concepts at second level was extended to develop classroom materials for teaching colours, gas behaviour and magnetism in primary schools. The last contribution is this part proposes that mathematics in modern school could be considered more as a "quasi-natural science" and asks how this approach may change teaching strategies.

Part II: Innovation in Undergraduate Physics Education presents teaching and learning of advanced physics topics based on and contributing to physics education research: the quantum Hall effect through the 5E model, a discussion of the foundations of thermodynamics, a carefully sequenced set of experimental activities concerning the photoelectric effect and quantum efficiency and the effect of introducing skills-oriented laboratories.

Part III: Trends in Physics Teacher Education explores how teacher education is changing to accommodate and promote different ways to teach and learn. A recurring theme is teachers assuming the role of their students as they develop an understanding of a variety of teaching approaches: they engage in role play about energy exchange, undertake inquiry into socio-scientific issues or experience a range of activities across the inquiry spectrum. Teachers reflect on learning sequences through critiquing them, and on classroom interactions through studying video recordings. The impact of teachers' views on the relation between mathematics and physics is also discussed.

Part IV: Bridging Gaps in Student Motivation and Engagement in Physics considers approaches that bridge the gap between formal, informal and non-formal learning environments and can lead to increased student motivation and engagement as well as increased awareness of careers and further programmes in physics.

It is our sincere hope that this collection of papers presents insights into current research in physics education at the end of 2017 and will be of interest to physics teachers, teacher educators and physics education researchers around the world with a commitment to bridging research and practice in physics teaching and learning.

The editors are grateful to the authors for their hard, fruitful work and to all the reviewers for their valuable remarks and time devoted to the development of the community of physics researchers and practitioners.

Dublin, Ireland   Eilish McLoughlin
                                                                                                            Paul van Kampen

Contents

Editors and Contributors

About the Editors

Eilish McLoughlin is an Associate Professor in the School of Physical Sciences and Director of the Research Centre for the Advancement of STEM Teaching and Learning (CASTeL) at Dublin City University. She received her Ph.D. in Experimental Surface Physics in 2000. Her research interests are focused on the development of curriculum, instruction, and assessment models in Physics/STEM education, at all levels of education, from primary school to Ph.D. level. She has led and collaborated on many research projects at EU, national, and local levels, including coordination of EU FP-7 project ESTABLISH and partner in EU projects 3DIPhE, OSOS and SAILS. She co-leads the physics initial teacher education programme at Dublin City University and has led several projects that support teachers, both in-service and pre-service, in adopting innovative practices in their classrooms and enhance student identity in Physics/STEM. She has received recognition for her leadership in STEM Education and engagement across formal and informal contexts: including the NAIRTL National Award for Teaching Excellence (2010); Institute of Physics Lise Meitner Medal (2018) and Institute of Physics Young Professional Physicist of the Year (2006) and Dublin City University's President Award for Public Engagement (2017) and for Teaching and Learning (2005).

Paul van Kampen is an Associate Professor in the School of Physical Sciences at Dublin City University and researcher within the School of Physical Sciences and the Centre for the Advancement of STEM Teaching and Learning (CASTeL) at Dublin City University. He obtained a Ph.D. in experimental atomic physics. His current research interests are divided between research-based development of teaching–learning sequences in university level physics and science teacher education. The focus in both areas is on how students develop scientific understanding and reasoning. He teaches physics and physics education at undergraduate and postgraduate level, and is heavily involved in science teacher education. His teaching is strongly influenced by his physics education research, which focuses on the development of research-led and research-validated educational materials. He was awarded the NAIRTL National Award for Excellence in Teaching Award for Excellence in Teaching and Learning in 2012. He chairs the Irish National Council for Curriculum and Assessment's Development Group for Junior Cycle Science.

Contributors

Antonio Amoroso holds a Ph.D. in Nuclear Physics at the University of Turin and is currently a research technician at the Department of Physics. He is member of the COMPASS collaboration at CERN since 2001, with research activities focused on spin physics and Drell-Yan process. He is member of the BES3 collaboration at IHEP since 2012, with research activities focused on CGEM (Cylindrical GEM) detector project. He also collaborates to the activities of the research group in history and didactics of physics, with a focus on the laboratory activities for secondary schools. He is the curator of the Museum of Physics of the University of Turin. For the academic year 2018–19, he will be teacher of the Physics module at the Physical education bachelor's degree and the Laboratory of Fundamentals and Didactics of Physics module of the Master's degree in primary education of the University of Turin.

Esther Bagno is a Senior Staff Scientist in the Science Teaching Department and has a major role in the research and development activities of the Physics Group. Her aim is to raise the quality of learning and teaching in Israeli high schools. Her work builds on her first-hand experience as a physics teacher for 17 years and own research. A major strand in her work has been the design, implementation and study of professional development frameworks for physics teachers aimed at nurturing teachers' professionalism (e.g. their knowledge of physics and physics teaching). She has been the director of the National Center for Physics Teachers since its establishment in 1995. She collaborates closely with the Israeli Ministry of Education and serves on several of its committees devoted to physics learning and teaching. She lectures and leads workshops and courses and serves on graduate programme committees.

Ami Baram is a psychodrama director and documentary film creator. His films and photographs were broadcast on the Israeli channel 2 and 10 and exhibited at the Israel museum, Jerusalem. He has varied experience in group directing—social groups, professional teams and therapeutic dynamic groups. He served as a teacher in the Kibbutzim College of Education Technology and Arts, Art high school and in a special programme for youth at risk. He has an ongoing interest in the interdisciplinary connection between education, documentation and group directing. He is the developer and co-director of "Video-didactic"—a project designed to develop physics and mathematics teachers' proficiency in conducting a peer video-based didactic discourse. He graduated from the Sam Spiegel Film School, Jerusalem, and the Kivunim Institute for Psychodrama and Group Work.

Sara Roberta Barbieri teaches mathematics and physics at high school. She graduated in physics with a thesis on mathematical models for the characterization of biophysical systems and she obtained a Ph.D. with a thesis on how to teach superconductivity at high school using mainly electromagnetic integral operators. From 2011 to 2017, she was in detachment to the physics department of the University of Milan, with which she is still collaborating. She conducts researches in the field of physics education in the context of electrical conduction, superconductivity and harmonic oscillations. She writes and illustrates stories to present physics in primary school and designs teacher training activities to introduce enquiry teaching at high school.

Leonardo Bellomonte is an Associate Professor of Solid State Physics (retired) at the University of Palermo. He has spent 2 years as a research associate at the University of Southern California, Los Angeles. He is a senior member of the American Physical Society (APS). His main field of research has covered paramagnetic resonance, lattice dynamics and optical absorption in solids. Other fields of research have concerned medical physics (mainly computer-based diagnosis of ocular states and/or pathologies); educational physics (devoted to the diffusion and implementation of computer-based teaching methods); computer simulation of various properties of matter such as the low temperature behaviour of the specific heat, etc.

Jan-Philipp Burde earned his Ph.D. in physics education research at Frankfurt University in Germany, where he developed and empirically evaluated a new curriculum to teach electricity in lower secondary schools. Prior to joining academia, he did his teacher training in the UK and gained qualified teacher status (QTS) to teach in schools in England. He also holds a graduate degree in Physics and English with a concentration in teaching and education from the University of Kassel, Germany. His research interests include the development of new teaching concepts, the use of computers in physics teaching and the effectiveness of context-based physics instruction.

Ruth Chadwick is a postdoctoral researcher and lecturer in initial teacher education. Her research interests include learning and teaching methodologies for science education, and curricular policy development. She completed her undergraduate degree in Zoology and subsequently obtained a PGCE in secondary education in science and biology. She worked for a number of years as a secondary school teacher in Scotland, during the introduction of the Curriculum for Excellence. From this, she gained an insight into curricular policy development from the "chalk-face". She completed a Ph.D. in 2018 with a thesis entitled "Development and Assessment of Scientific Literacy for Secondary Level Science

Education". The focus of this research was on curricular policy development and assessment in the science curricula of Scotland and Ireland in both secondary and tertiary level education.

Federico Corni is Professor of Didactics of Physics at the University of Modena and Reggio Emilia. From February 2019, he will take up a position in the Faculty of Education at the Free University of Bolzano. He teaches physics to prospective kindergarten and primary school teachers and trains groups of in-service teachers in various schools. His research focuses on conceptual metaphor in building and assessing scientific concepts and narrative approaches to physics and science teaching. He is author of journal papers and editor of "Le scienze nella prima educazione. Un approccio narrativo a un curricolo interdisciplinare" (Erickson, Italy, 2013).

Anna De Ambrosis is Associate Professor in Physics at the Physics Department of the University of Pavia (Sector FIS/08 Physics Education and History of Physics). From 1988 to 1991, she worked at the University of Ferrara and since 1991 at the University of Pavia. She served as coordinator, for Mathematics and Physics, of the Postgraduate School for Teachers' Preparation at the University of Pavia and for Physics in the Postgraduate Teacher Training Course (TFA). Her scientific interests lie in Physics Education. Research in this field has been developed along various lines: the role of computer simulations, correlated to students' experimental activity, in the formalization process; the use of MBL tools and new technologies to favour the teaching/learning of physics at different school levels; the introduction of topics of modern physics, in particular special relativity and quantum physics, in the high school curriculum; the design and implementation of models and strategies for secondary school physics teachers preparation. The research has been carried out within national and international projects.

Claudia De Grandi recently joined the Physics and Astronomy Department at University of Utah as an Assistant Professor of Educational Practice. One of her main roles is to implement and promote evidence-based teaching pedagogies in introductory physics courses and improve the teaching quality and inclusiveness of physics and STEM courses more broadly. She works closely with Learning Assistants and Teaching Assistants to create effective instructional teams. At University of Utah, as well as previously as a Teaching Postdoc at Yale University, she has been working to reform the Introductory Physics curriculum for Life Sciences majors (pre-medical students and biology majors), both the lecture courses, as well as the laboratory ones. Her background is in condensed matter theory; she obtained her Ph.D. from Boston University studying quantum systems of cold atoms, and then pursued research on quantum information during her postdoctoral time at Yale University.

Bat-Sheva Eylon is a faculty member of the Science Teaching Department at the Weizmann Institute of Science (WIS), and acted as its head in 2008–2015. She is a Fellow of the AAAS, recipient of the Israeli EMET Prize in Education for 2015 and pedagogic head of the Rothschild-Weizmann M.Sc. Program for Excellence in Science Teaching. She studies the learning and teaching of physics and physical sciences in secondary schools and a major focus has been the long-term professional development of teachers in professional learning communities, presently implemented in grades 7–12 all over Israel. She co-authored with Linn the book: *Science learning and instruction: Taking advantage of technology to promote knowledge integration* (2011). She holds a B.Sc. in physics and mathematics from the Hebrew University of Jerusalem, an M.Sc. in physics from the WIS and a Ph.D. in science education from Berkeley University in the US.

Odilla E. Finlayson is Associate Professor of Science Education in the School of Chemical Sciences, Dublin City University (DCU). She is one of the founding members of CASTeL (Centre for the Advancement of STEM Teaching and Learning) at DCU and plays an active part in its management and development. She is involved in teaching chemistry to undergraduate students and pre-service teachers. Current research interests are in sustaining science across transitions and, in particular, in development of appropriate science (chemistry) curricula and assessment. She is actively involved in EU projects such as ESTABLISH and coordinator of SAILS.

Enrico Giliberti is a researcher in Didactics and Special Education at the University of Modena and Reggio Emilia. He teaches education technologies and group work methodology in a primary school teacher degree course; didactics, special education and education technologies in an educator degree course. He trains groups of in-service teachers in the subjects of science education and playing/table gaming. His research focus is on conceptual metaphor in building and assessing scientific concepts and narrative approaches to science teaching in primary school and the role of playing as a tool for knowledge. He is a member of the scientific committee of the Research Centre for Metaphor and Narrative in Science (www.manis.unimore.it).

Marco Giliberti graduated in physics at the University of Milan with a thesis on the foundations of quantum physics. He is a researcher at the same university where he teaches "Preparation of Teaching Experiences 1 and 2" and "Relativity". He is responsible for the Scientific Degree Plan (Physics) at unimi. He obtained the national scientific qualification to associate professorship in 2017. He performs research in quantum physics education, in physics teaching through theatre and in teacher training. He is the author of about 60 publications in national and international journals, the book "Physics at theater"(Aracne editrice, 2014) and the monography "Peter Higgs" (Grandangolo Scienza–Corriere della sera, March

2017). He is co-founder of the group "The Physics' Show" and co-author and actor of seven theatrical performances of physics that, so far, have had about 400 performances in Italy and abroad.

Mark D. Greenman is a Research Fellow in the Department of Physics College of Arts and Science, Boston University. He serves as co-PI and director of Project Accelerate NSF#1720914 and is co-PI on Project PSUNS Robert Noyce Teacher Scholarship Program NSF#1660681. He is a past Presidential Awardee for Excellence in Science Teaching, Janet Guernsey Awardee for Excellence in Physics Teaching, and served 2 years as an Einstein Fellow at the National Science Foundation Division of Undergraduate Education. His undergraduate physics degree is from Hofstra University and graduate physics degree from Syracuse University. A passion for physics and physics teaching and learning are hallmarks of his career.

Claudia Haagen is Full Professor at the institute of Physics at the University of Graz where she is the department's Chair for Physics Education and deputy director of the institute. Claudia Haagen has 8 years of experience as high school teacher. She earned her Ph.D. in Educational Sciences at the University of Graz. In 2016, she completed her habilitation in physics education and gained the venia docendi for didactics of physics at the University of Vienna. In 2017, she was awarded the Josef-Krainer-Würdigungspreis—a highly competitive science prize—for her habilitation, a design-based research project on teaching and learning introductory optics for year 8. Professor Haagen's research interests lie in the area of content specific teaching and learning processes and conceptual change, professional development of teachers and language in science teaching.

André Heck earned M.Sc. degrees in mathematics and chemistry, and a doctoral degree in mathematics and science education. He is a senior lecturer at the Faculty of Science of the University of Amsterdam, teaching in particular mathematics and physics to life science students. His main research area is the application of ICT in mathematics and science education.

Martin Hopf is Full Professor for Physics Education Research at the University of Vienna in the Austrian Educational Competence Centre Physics. He studied mathematics and physics at the LMU Munich and taught in high schools for 5 years. He earned his Ph.D. in the PER group at the LMU Munich. His main research interests are students' conceptions and didactical reconstruction. He was awarded the Polytechnik-price in 2011 for the 2DD-project, in which a reconstruction of Newtonian physics for 7th grade students was developed and evaluated. It turned out that this curriculum worked significantly better than traditional curricula.

Andrea Kárpáti is Professor and UNESCO Chairholder at Eötvös Loránd University, Faculty of Science, Centre for Science Communication and Multimedia in Education. She is Head of the Visual Culture Research Group of the Hungarian Academy of Science and ELTE University. Her teaching and research activities involve visual culture of children and adolescents, digital literacy in education, museum learning and synergies of science and arts education. Her recent research projects include EU-funded EnVIL (developing the European Framework of Visual Literacy), Moholy-Nagy Visual Modules (a curriculum innovation project) PARRISE (representing socially sensitive issues in science communication) and KP-Lab (developing collaborative pedagogical scenarios).

Andrea Király is a physicist and a teacher of mathematics and physics. She obtained her Ph.D. in Statistical Physics with a thesis entitled "Long-range correlations in daily temperature records". She is Assistant Professor at ELTE Eötvös Loránd University, Faculty of Science, Centre for Science Communication and Multimedia in Education. She is head of the Science Centres and Informal Learning Working Group of the MTA-ELTE Physics Education Research Group established by the Eötvös Loránd University and the Hungarian Academy of Sciences in 2016, and was head of the Hungarian team in the PARRISE (Promoting Attainment of Responsible Research and Innovation in Science Education, 2014–2017) EU FP7 project.

Inkeri Kontro obtained her Ph.D. in the field of Materials Physics in 2016 from the University of Helsinki, where she currently works. She has also been a visiting researcher at the Niels Bohr Institute, Copenhagen. During her Ph.D., she started working on the development of introductory and intermediate physics courses. Her research interests in Physics Education Research are the development of content knowledge and its interplay with student attitudes towards physics.

Eduard Krause studied mathematics and physics at the University of Siegen (Germany). In 2013, he received his doctorate in the field of physics education research. He was appointed as Assistant Professor at the Institute of Mathematics education research in Siegen in 2014. In 2015, he was a guest lecturer at the Hanoi National University of Education (Vietnam). In 2016, he held a professorship at the University of Cologne and in 2018 at the Friedrich Schiller University in Jena. His research interest lies in the didactic implications resulting from the synergies between mathematics and physics. In this area, he successfully leads international projects.

Alessandra Landini, Ph.D. student in Human Sciences and Didactic of Physics at University of Modena and Reggio Emilia, is a primary school teacher and teacher educator and collaborates with the Centre for Metaphor and Narrative in Science. Her research includes the use of narrative, metaphorical expressions and gestures in science education, the embodied cognition related to didactic methodologies from kindergarten to secondary school and inclusive pedagogy.

Yaron Lehavi is a physics teacher, a senior lecturer, the head of the Research and Evaluation Authority at the David Yellin Academic College of Education and head of the National Center for Physics Teachers at the Weizmann Institute of Science in Israel. He has a long experience in teachers' training, development of learning materials, curriculum development and scientific and educational counseling. He is the developer and co-director of "Video-didactic"—a project designed to develop physics and mathematics teachers' proficiency in conducting a peer video-based didactic discourse. His research covers such areas of physics education (at various levels) as conceptual understanding of physical concepts, teachers' PD and the Phys-Math interplay. At the beginning of his career, he served as a high-school physics and mathematics teacher. He holds a B.Sc. in physics and mathematics, an M.Sc. in physics and a Ph.D. in science education from the Hebrew University of Jerusalem.

Matteo Leone is Associate Professor of didactics and history of physics and during 2016–19, Dean of the University Program for pre-primary and primary school teaching at the University of Turin. Since 2016 he is Director of the Museum of Physics belonging to the University of Turin Museum System. His research activities focus on the nineteenth and twentieth centuries' history of physics through the recovery and analysis of archival documents and other primary sources, on the physics education at the primary school level and on the relationships between history of physics and physics education. At present, he is supervisor of a Ph.D. research project in didactics and history of physics devoted to the census of the collections of demonstration instruments of the classic lyceums physics cabinets in Piedmont and to the exploration of the educational significance of these collections.

Massimiliano Malgieri is a postdoctoral researcher in the field of Physics Education at the University of Pavia. He graduated in Physics in 2003 at the University of Genova, where he specialized in teaching physics and mathematics in high school. He then worked for several years as a school teacher before returning to University to obtain his Ph.D. In 2015, at the University of Pavia, he successfully defended his Ph.D. thesis, titled "Teaching quantum physics at introductory level: a sum over paths approach". Since 2015, he is working in Pavia as a postdoctoral researcher and contract professor for courses such as Physics Education and Preparation of Educational Experiences. He is mainly involved in research and projects related to improving the teaching of modern physics at secondary school level, designing and disseminating innovative teaching–learning sequences and laboratory experiments, exploring the role of games and simulations in the learning process. He is the author of more than 20 articles in international journals indexed by ISI or Scopus, and the co-author (with Ugo Besson) of a book on the teaching of modern physics in secondary school.

Daniela Marocchi was Associate Professor of Experimental Physics at the University of Turin from 1985 to 2016. She had been President of the Master Degree in Physics, University of Turin responsible of the National Science Degree Project (PLS) since the beginning of the project, and referent for the Physics qualification courses for secondary school teachers. She encouraged the research in physics education at the Department of Physics of the University of Turin by proposing a master degree course more fitting to the needs of future teachers. She was supervisor of several master's theses in physics education and she is currently collaborating to a Ph.D. project in History and Didactics of Physics. For the academic year 2018–19, she will be teacher of a module on Educational Methodologies and Technologies in Physics at the Master Degree in Physics.

Avraham Merzel is a lecturer (equivalent to Assistant Professor) at the School of Education at the Hebrew University of Jerusalem, and a high-school physics teacher. He teaches pre-service physics teachers and leads a professional community of in-service physics teachers. His research aims to develop the capacity to mentor colleagues in teachers, and to study concepts and methods in science teaching. He participates in the project: "Video-didactic"—a project designed to develop physics and mathematics teachers' proficiency in conducting a peer video-based didactic discourse. He holds a B.Sc. in physics and cognitive sciences, an M.Sc. and a Ph.D. in cognitive science from the Hebrew University of Jerusalem.

Simon G. J. Mochrie is a Professor of Physics and of Applied Physics. He has undergraduate and graduate degrees in Physics from the University of Oxford and the Massachusetts Institute of Technology, respectively. Currently, he carries out experimental and theoretical research in biological physics at Yale University, focusing on the organization and dynamics of chromatin. Over the last several years, in collaboration with Claudia De Grandi and Rona Ramos, he has developed and taught a re-imagined version of Introductory Physics for the Life Sciences (IPLS), that eschews the traditional syllabus and instead emphasizes topics that are biologically and medically meaningful, such as diffusion, viscous fluid flow, mathematical modeling and statistical mechanics, while at the same time seeking to make the class as accessible and inclusive as possible.

Pasquale Onorato is associate professor at the University of Trento. He graduated in Physics in 1998 at the University of Napoli where in 2001, he successfully defended his Ph.D. thesis. In 2001, he moved to the Frascati National Laboratory of the INFN. Then, he was postdoctoral researcher in the Physics Department of the University of Pavia, Italy, working with the Physics Education Research group. Since 2015, he works at the Department of Physics, University of Trento in the Laboratory of Physical Science Communication. He is author or co-author of more than 90 scientific papers in international ISI-Scopus indexed journals and of several contributions to national and international conferences. He is actively involved in investigations about physics education, and the use of technology to enhance learning physics, and was also interested in researches on theoretical condensed matter. His main research interests are students' conceptions and reasoning; design and experimentation of innovative teaching–learning sequences; modern physics instruction; strongly correlated electron systems in low dimensions; carbon-based nanostructures (nanotubes, graphene, etc.); spin effects in low dimensional electron systems: spin Hall effect and spintronics, and Raman spectroscopy.

Dominique Persano Adorno holds Ph.D. in Applied Physics, Assistant Professor of Solid State Physics and is a member of the Physics Education Research Group at the Department of Physics and Chemistry of Palermo University. Her research interests include the study of nonlinear dynamics of complex systems in condensed matter physics and applied physics (spin and charge transport phenomena, 2D materials and graphene, polymer translocation, cell growth, etc.). Recent investigations deal with the design and implementation of inquiry-based learning environments for the development of effective strategies to teach physics at school and university level. She is involved in research projects concerning Inquiry-based Science Education, innovative e-learning scenarios, collaborative learning, ICT for enhanced learning. She is the author of more than 60 papers in ISI/Scopus indexed journals; 5 book chapters; over 50 refereed proceedings to international conferences. She is a member of the Editorial Board of PLOS ONE, Cogent Physics, International Journal of Biomedical Engineering and Science, Applied Physics Research.

Nicola Pizzolato is a physics teacher at secondary school and researcher focused on applied physics in Palermo, Italy. He earned an M.Sc. in physics (astrophysics) in 1997 from the University of Palermo. He was a visiting scientist from September 1999 to March 2000 at the Harvard-Smithsonian Center for Astrophysics-Cambridge, USA, to study the calibration data from the CHANDRA X-ray space telescope. He received a Ph.D. in physics from the University of Palermo in 2002 and worked as a postdoctoral fellow at the Palermo Astronomical Observatory. More recently, he focused his work on the field of Physics Education Research. He earned a second Ph.D. in Physics Education in 2014, defending a thesis about the development of effective strategies of teaching–learning science in upper secondary school and at university level, by following an inquiry-based approach to physics education. He has published more than 80 scientific publications, including about 40 papers in ISI journals.

Gesche Pospiech is Full Professor of physics education at the Faculty of Physics of the Technische Universität Dresden in Germany. Her main research interest concerns Modern Physics in secondary school with a focus on quantum theory with a modern approach. Her second field of research is mathematics in physics education with emphasis on secondary school. In addition, she studies the pedagogical content knowledge of teachers in this field. She has been part of European projects fostering the science interest of students and in the Network Particle world in Germany. In addition, she is responsible for the physics education training of teacher students at TU Dresden.

Stefan Radl holds a Ph.D. in chemical engineering from Graz University of Technology (TU Graz) and was promoted with highest distinction (promotio sub auspiciis Praesidentis rei publicae) in 2011. The topic of his Ph.D. was "Modeling of multiphase systems in pharmaceutical applications", the thesis was supervised by Prof. Johannes Khinast. Prior to this, he received a Diploma (Master of Science) in chemical engineering from TU Graz, with a thesis on the modeling of bubbly flows. Since 1 August 2018, Stefan is an Associate Professor at TU Graz, where he holds a venia docendi for the subject "particle technology". Prior to that, Stefan was a Post Doc at Princeton University (Deparment of Chemical and Biological Engineering, supervisor: Prof. Sankaran Sundaresan, 2011 to 2012), and an Assistant Professor at TU Graz (from 2012 to 2018).

Rona Ramos is a Lecturer and instructional lab manager in the department of physics at Yale University. She holds a B.S. in physics from UC, Berkeley, a masters in physics from UCLA and a Ph.D. in applied physics from Yale University where she developed techniques for the MRI of solids. Dr. Ramos teaches introductory physics courses and is also involved in training postdoctoral and graduate students in active learning and inclusive teaching methods based on cognitive science and physics education research. She is also co-director of Girls Science Investigations, a Yale outreach programme encouraging middle school girls to pursue careers in science.

Jakob D. Redlinger-Pohn studied Chemical and Process Engineering at Graz University of Technology. His diploma work was related to the drying of dense and wet particle slurries. In his Ph.D. work, which was defended 2018, he studied the size dependent clustering of particles (here cellulose fibres) and selective removal of smaller (here shorter) particles. During his Ph.D. work, he lectured class and lab courses on the handling, description and manipulation of particles and particle assemblies.

Erich Reichel received a diploma for teaching physics and mathematics at grammar schools at the University of Graz. Afterwards, he was Assistant Professor at the Institute for Experimental Physics in the field of lasers in medicine and was promoted in biophysics (Ph.D., 1984). From 1992 to 2013, he was a high school teacher of physics and mathematics at the BG/BRG Seebachergasse in Graz. In parallel, he was a partner in national and international school projects (Comenius, FP7) and was Lecturer in physics at the University of Graz and the University College of Technology (FH Joanneum, Graz). During this time, he was also responsible for Styrian physics teachers and their in-service education. In 2013, he moved to the University of Teacher Education of Styria as a university college professor for didactics in science and technology. He is currently setting up NATech (Center for Didactic Research in Science Education).

Marta Rinaudo is Ph.D. student in Physics and Astrophysics, at the University of Turin (Italy), with a project on the history of physics and physics education titled "Towards an integrated Museum of Physics in Piedmont". After graduating, she planned and implemented hands-on laboratory activities for physics education at primary and secondary schools and participated in the e-learning project offered by the Department of Physics. She had been teacher at the Laboratory of Fundamentals and Didactics of Physics module of the Master degree in primary education of the University of Turin. As part of her Ph.D. project, her current research activities are carried out at the Museum of Physics of the University of Turin, with the main focus on studying the origins of the old Physics Cabinet and developing educational activities based on the collection of historical instruments. Also, she is actively working on the establishment of a regional network of school museums of physics with the goal of discovering the educational value of the historical scientific heritage.

David Sands studied Applied Physics at the University of Bradford, graduating in 1982 with a First Class Honours degree. He stayed on at Bradford to research electronic characterization of semiconductor interfaces, achieving his Ph.D. in 1987. He moved to the University of Hull in 1990 and continued to work on electrical characterization of semiconductors for a while, but gradually concentrated more on laser processing. Whilst at Hull he also developed a strong interest in physics education and in September 2004 was awarded a University Teaching Fellowship. This was a competitive award designed to mirror the National Teaching Fellowship at that time. David was awarded his NTF in 2015. He is currently Director of Learning and Teaching for the School of Mathematics and Physical Sciences at Hull. He chairs the Degree Accreditation Committee of the IOP, the IOP Higher Education Group and the Physics Education Division of the European Physical Society and also represents the UK on Commission C14 of IUPAP (ICPE). As well as his interest in physics education, he has pursued a strong interest in the foundations of thermodynamics and continues to work in this field today.

Ruti Segal is a mathematics teacher educator and served as a high-school teacher of mathematics for 26 years. She also served as an assistant to the national superintendent of high-school mathematics in Israel. She is a Lecturer at Oranim College of Education and at Shaanan College of Education. She integrates technology in mathematics instruction and research. Her research focuses on professional development of mathematics teachers and mathematics teacher educators. She serves as an academic advisor at the Weizmann Institute for "Video-didactic"—a project designed to develop physics and mathematics teachers' proficiency in conducting a peer video-based didactic discourse. She holds a B.Sc. in mathematics and computer sciences from the Haifa University, an M.Sc. in teaching science focus in mathematics from the Jerusalem University-Givat Ram, and a Ph.D. in teaching science focus in mathematics from the Technion - Israel Institute of Technology.

Verena Spatz is a Junior Professor at the Institute of Physics at the TU Darmstadt. After she had completed her studies of mathematics and physics with honours, she earned her Ph.D. in physics education research at the LMU Munich in 2010. Over the next 5 years, she was a teacher at several German high schools and an external lecturer at the University of Bamberg. Then she spent one year as a postdoctoral researcher at the University of Vienna in the Austrian Educational Competence Centre Physics. She was awarded the Science Prize (University of Augsburg) in 2008 and the Polytechnic Prize (Polytechnic Society of Frankfurt) in 2011. Her main research interests are processes of conceptual change, implementation of instructional innovation as well as effects of mindsets on science teaching and learning.

Warren B. Stannard is presently working within the Australian Education system to assist high schools introduce technology programmes into the curriculum. He has a Ph.D. in Physics (materials analysis by nuclear techniques) and a Masters by research in a similar field. Recently, he completed a project at the University of Western Australia to evaluate student responses to the introduction of Einsteinian Physics early in their educational life. He has worked as the science education officer at the Gravity Discovery Centre in Gingin, Western Australia. He has previously worked as a Lecturer in Physics at Edith Cowan University and taught at Deakin University, delivering lectures in Engineering Physics, Machine Dynamics, Principles of Thermodynamics, Materials Science and Engineering Design.

Péter Tasnádi is a retired professor of the Eötvös Loránd University, Faculty of Science. He was the Deputy Dean of Education Affairs of the Faculty of Science for 10 years and the President of the Centre of the Science Education Methodology of the Faculty of Science. His teaching activity involves introductory physics courses, lectures on teaching methodology and laboratory courses in teacher training, as well as dynamic meteorology and atmospheric physics for graduate and postgraduate students in meteorology. His research field is the methodology of physics teaching, metal physics and atmospheric physics. Besides research papers, he has written a number of textbooks for secondary school and university students.

Onne van Buuren is a physics teacher at a Dutch school for secondary Montessori education, a developer of instructional materials for secondary physics education and an educational researcher. In 2014, he obtained his doctoral degree in Physics Education at the University of Amsterdam, with a thesis on the development of a modelling learning path for lower secondary physics education. Since 2018, he is also a physics teacher trainer at the Vrije Universiteit Amsterdam.

Thomas Wilhelm graduated from grammar school in 1989, studied at the University of Würzburg and then completed the practice teaching period. After this, he worked for several years as a teacher at grammar school. This was followed in 2000 by a time as a research assistant at the University of Würzburg (with the conferral of a doctorate in 2005). In 2011, the habilitation was completed. In 2011, Prof. Wilhelm accepted a professorship at the University of Augsburg and in 2012 a professorship at the University of Frankfurt, where he has been the Executive Director of the Department for Physics Education Research since 2014.

Part I
Development of Physics Teaching and Learning in School

Using the Electron Gas Model in Lower Secondary Schools—A Binational Design-Based Research Project

Claudia Haagen⊙, **Jan-Philipp Burde**⊙, **Martin Hopf**⊙, **Verena Spatz**⊙ and **Thomas Wilhelm**⊙

Abstract Students' understanding of introductory electricity concepts can be fragmented even after instruction. Several reasons for this have been identified: electricity is an abstract and complex topic and traditional instruction frequently fails to meet students' learning needs. As a consequence, conceptual change is not triggered and misconceptions prevail. Research findings show that the concept of voltage, in particular, presents many difficulties for students. This paper presents a research project from four working groups whom are collaborating to develop a teaching strategy for introductory electricity at lower secondary schools which is based on the electron gas model. Additionally, research finding from one project partners' implementation of this teaching approach in a pilot study are reported. The concept of voltage is introduced as an electric pressure difference across a resistor in an electric circuit. The evaluation of this approach with more than 700 high school students shows very promising results. Based on these findings, a Design-Based Research project has been jointly developed between two German and two Austrian Universities. The aim of this study is to find out whether the significantly better performance of students instructed according to the electron gas model can be replicated with a wider sample of teachers and students across the project partners' locations.

C. Haagen (✉)
University of Graz, Universitätsplatz 5, 8010 Graz, Austria
e-mail: claudia.haagen@uni-graz.at

J.-P. Burde · T. Wilhelm
Goethe-University Frankfurt, Max-von-Laue-Str. 1, 60438 Frankfurt am Main, Germany
e-mail: burde@physik.uni-frankfurt.de

T. Wilhelm
e-mail: wilhelm@physik.uni-frankfurt.de

M. Hopf
University of Vienna, Porzellangasse 4/2/2, 1090 Vienna, Austria
e-mail: martin.hopf@univie.ac.at

V. Spatz
Technical University of Darmstadt, Hochschulstraße 12, 64289 Darmstadt, Germany
e-mail: Verena.Spatz@physik.tu-darmstadt.de

© Springer Nature Switzerland AG 2019
E. McLoughlin and P. van Kampen (eds.), *Concepts, Strategies and Models to Enhance Physics Teaching and Learning*,
https://doi.org/10.1007/978-3-030-18137-6_1

Keywords Electricity · Design-Based research

Introduction

Electricity is one of the major topics of the physics curriculum in secondary schools. Electricity and electronic devices are essential parts of our daily life. Nowadays, we can hardly imagine a world without electronic devices. In our daily routines, we have to make decisions which require at least a minimal knowledge about electricity: What voltage batteries do we need for the TV remote control? Which charging unit supplies the appropriate voltage for charging our mobile phone? Why is it dangerous to put a fork into the toaster to get a slice of toast out of it?

Moreover, it is of course important to know that many processes in the human body are based on electrical impulses and signals. Additionally, there are social issues related to electricity, such as, sustainable consumption or alternative energy sources etc., which requires a level of scientific literacy in this field. Thus, a basic understanding of electricity is fundamental for our daily private and professional life as well as for active citizenship.

A basic understanding of electricity includes, above all, a correct conceptual understanding of simple electric circuits and the key concepts of current, voltage and resistance. Although electricity is a main topic in lower and upper secondary schools, many students develop only a fragmented understanding of the basic physical concepts of electric circuits during secondary education. This phenomenon is not particularly new or limited to a certain age group. McDermott and van Zee [1] observed that, even university students, frequently interchanged the words current, energy, power, voltage, and even electricity when talking about circuits.

Research findings also highlight the idea that current is consumed in a circuit is still maintained by students even after instruction [2]. Another major challenge is the concept of voltage. It is particularly difficult for students to distinguish between electric current and voltage. Frequently, students only operate with the concepts of current and resistance and consider voltage as a property of electric current. In general, the concept of electric current seems to determine students' understanding of electric circuits. As a consequence, it is not necessary for them to conceptualize voltage as an independent physical quantity.

Studies on students' learning and conceptions in introductory electricity have been given considerable attention in the past three decades. Insights about domain specific learning have, however, rarely affected school teaching practice, since knowing students' alternative conceptions does not necessarily mean that educators have practical methods to promote students' conceptual change. As conceptual change is often not triggered by conventional instruction, misconceptions prevail. Subsequently, students' understanding of introductory electricity remains in many cases fragmented, even after instruction. Several reasons can be identified for this: electricity is an abstract and complex topic and traditional instruction frequently fails to meet students' learning needs. The design of successful learning environments needs to merge subject matter knowledge with student perspectives (contexts, interests, etc.) and learning needs.

Motivation

We find, at least in German speaking countries, that reviewing the content of school textbooks reveals that the introduction of electricity is mainly based on the concepts of current, resistance and voltage, while the concepts of potential and potential difference are hardly addressed [3]. From the perspective of subject matter knowledge, voltage refers to a potential difference by definition. In conventional educational reconstructions of the topic electricity, the key idea of voltage as a potential difference is, however, hardly ever included. Taking students' learning paths into account, it seems to be quite likely that a better understanding of the concept of voltage requires an understanding of the concept of electric potential first [4]. In physics education research, numerous studies have concluded that the introduction of voltage as potential difference supports students' learning needs [5–7].

Examples of such approaches include the "*flat water circuit analogy with a double water* column" as shown in Fig. 1 [8, p. 35] and the 'stick model' developed in Munich [7] as illustrated in Fig. 2. In both teaching concepts, "the visual representation of the potential has proven to be an important factor for the learning success as it facilitates the build-up of a mental model of the electric potential [7, p. 70; 8, p. 35]" [3, p. 27].

Based on these findings, which support the introduction of voltage as potential difference, Burde and Wilhelm [3] have developed a teaching approach for introductory electricity which uses a content structure based on the electron gas model. In an intervention study with a control group of students, the teaching approach based on the electron gas model was shown to lead to significantly better learning outcomes for the intervention group [10].

Fig. 1 The flat water circuit analogy: a double water column visualises the potential difference

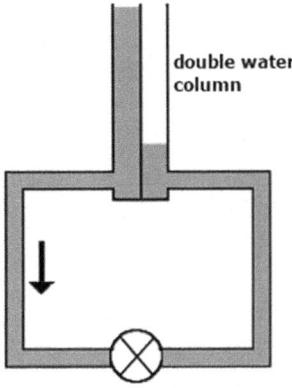

double water column

Fig. 2 The stick model [9, p. 122]: sticks positioned in different heights above the base visualize the electric potential difference

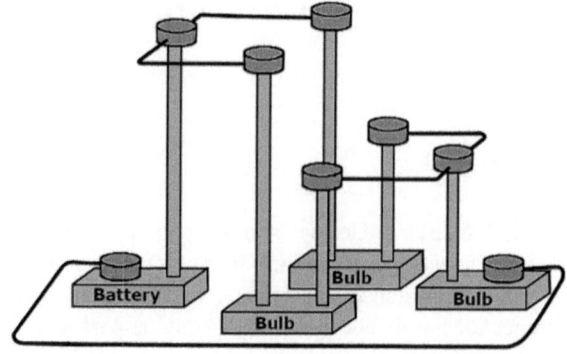

Based on these findings, a Design-Based Research (DBR) project has been jointly developed by two German and two Austrian universities. The aim of this study is to examine whether the significantly better performance of the control group of students instructed with the electron gas model can be replicated with another sample of teachers and students at the project partners' locations. In addition, it is of interest if variations concerning contexts used can stimulate students' interest.

The Electron Gas Model—Previous Findings

It is well known that students base their understanding of simple electric circuits on the concepts of current and resistance [11] and thus fail to develop a solid conceptual understanding of the concept of voltage. However, previous research studies suggest that the explicit introduction of voltage as potential difference represents a successful approach. This idea was exploited by Burde and Wilhelm [12, 13] who developed a teaching approach, which is based on the electron gas model: Building on students' prior knowledge about air pressure, voltage is introduced as an electric pressure difference across a resistor in an electric circuit. The intention of the teaching approach based on the electron gas model is to facilitate a qualitative and solid concept of electric potential so that students are able to analyze simple circuits appropriately. In terms of learning theories, this approach builds on diSessa's "Knowledge in Pieces" theory [14, 15] which stimulates conceptual development by building on students' prior knowledge. In this study, students' intuitive understanding of air pressure serves as a qualitative model of the electric potential in conducting wires.

Burde and Wilhelm specify the analogy between air pressure and voltage as follows:

> It is important to note that we are not talking about the physical pressure concept in the technical sense of a state variable here, but an intuitive prototype concept of 'pressure' in the sense that compressed air tries to push itself out of a container, e.g. based on everyday life experiences with air pumps, bicycle tires or air mattresses. [3]

The air analogy is more compelling to learners than the water analogy, mainly because of the palpable compressibility of air and the fact that students gather experiences with the compressibility of air in everyday life (e.g. bicycle tires or air mattresses). In contrast, water is often perceived by students as an incompressible fluid and they have hardly any experience with water pressure from everyday life. Because water under high pressure differs neither visibly, nor palpably, from water under low pressure, the water analogy has proven to be not as helpful as commonly assumed. A more detailed discussion of the advantages and disadvantages of the water circuit analogy can be found in [10, 16].

It is quite intuitive for students that differences in air pressure cause an air flow. As an analogy, the electron gas model is used to introduce voltage as potential difference, which can be understood as electric pressure difference across a resistor. So, it is quite plausible that similarly to air pressure differences, which cause an air flow, electric pressure differences (electric voltage) cause an electric current as shown in Fig. 3.

The teaching materials based on the electron gas model were tested in a study with treatment and control group classes in schools nearby to Frankfurt (Germany) using a pre- and post-test design. For the participating teachers in this study, the intervention consisted of their participation in a short afternoon training session where the teaching materials were introduced and distributed.

The evaluation of this approach using the electron gas model with year 7 and year 8 students (N = 790) showed that the students of the treatment group not only significantly outperformed the conventionally instructed students, but also had a better conceptual understanding of voltage (see Fig. 4). A more detailed analysis of student's tests revealed that girls, especially within the treatment group, showed a smaller conceptual gain than boys, although their pre-knowledge was not

Fig. 3 The electron gas model [17]

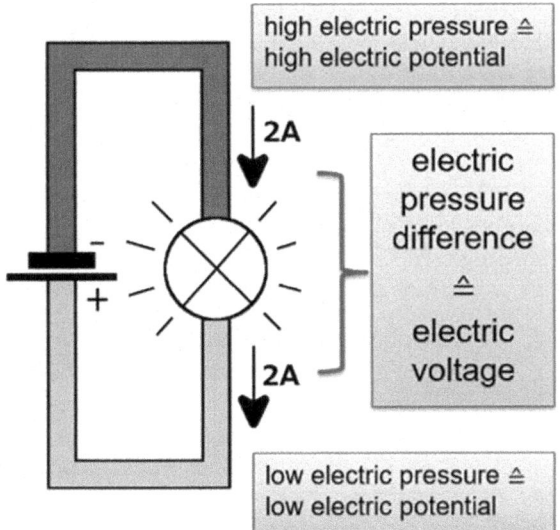

Fig. 4 Pre-post results of the control and treatment group [17]

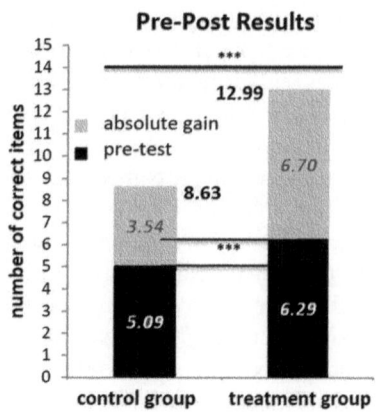

significantly different. These results are the starting point for our bi-national DBR project, which builds on the content structure of Burde and Wilhelm [3, 10, 13] and their findings.

Goals and Research Questions

The goal of our bi-national Design-Based Research (DBR) project EPo/EKo ("Electricity with Potential/Electricity with Contexts") is to refine the content structure developed by Burde and Wilhelm [3], which proved a promotion of deeper conceptual understanding than conventional instruction. Within the framework of Design-Based Research [18, 19] the project partners aim to implement research findings in domain specific learning into school practice by offering teachers research based teaching materials and short in-service training workshops. At the same time, this project aims to gain deeper insight into students' domain specific reasoning in introductory electricity with a special focus on the concept of voltage. Another goal of the bi-national project is to investigate the effects of the integration of "interesting" contexts (e.g. ROSE study [20]) in the teaching concept on students' interest, self-concept and conceptual understanding. Additionally, participating teachers' pedagogical content knowledge (PCK) and its development will be investigated.

The research questions of this DBR study can be summarized as:

- Can the significantly better performance of students instructed with the electron gas model be replicated with another sample of teachers and students at project partners' locations?
- What are the effects of integrating "interesting" contexts within the teaching concept on students' interest, self-concept and conceptual understanding?

• What pedagogical content knowledge (PCK) and beliefs about teaching and learning introductory electricity do participating teachers have and how do they change due to the implementation of new teaching materials in their classes?

Research Design

The project is based on a pre-post-follow-up-design with one control and, in total, three intervention groups. Figure 5 shows the treatments for the control group and the intervention group, which consist of three different cohorts across three school years.

The group of the participating teachers (T) is the same for each year. They teach physics in a 7th grade in three subsequent years. In the first year of the project, the participating teachers (T) follow their usual approach to introductory electricity. Pre-post-follow-up-tests are administered to set a baseline for the students' performance of the cohort of year one (StY1) as can be seen in Fig. 6. A test, which focuses on the teachers' PCK facets and beliefs concerning introductory electricity is administered after the instruction. In addition, case studies are conducted with some teachers using interviews. At the end of the first year, teachers take part in a short workshop in preparation for the intervention studies which take place in the second year.

In the second year, teachers are split in two different intervention groups (T1&T2) (see Fig. 5): T1 teach their cohort of year two (StY2) focusing on context orientation and follow their usual content structure. T2 teach their cohort of year two (StY2) following the content structure of the electron gas model without focusing on context orientation. Figure 7 illustrates the research design of year two, where a pre-post-follow-up-design is used with students. This time, all student groups (StY2) function as intervention groups, revealing either the influence of the electron gas content structure or the influence of context orientation. The PCK test is again administered to teachers after they complete year two instruction. The teachers at the centre of

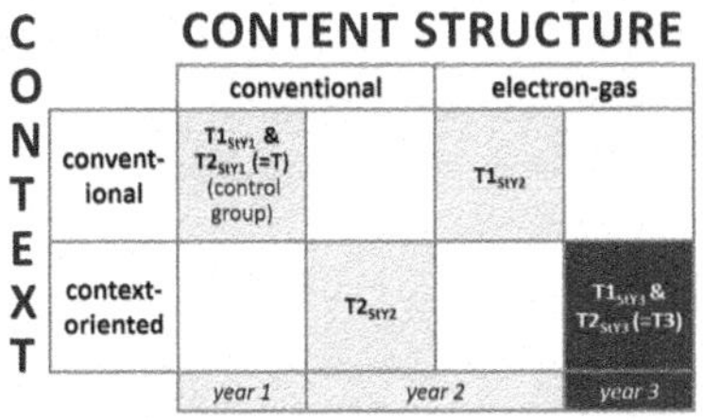

Fig. 5 Control and intervention groups in the EPo/EKo ("Electricity with Potential, Electricity with Contexts") project. T stands for the participating teachers. T1 is the group of teachers who teach the electron gas course in year 2 (EPo) and T2 is the group of teachers who teach the context-oriented course in year 2 (EKo). StY identifies the students (St) of each project year (Y). T1StY2, for example, refers to the group of students who are in the context-oriented course (T1) in the second year of the project (StY2)

the case studies are interviewed again. Finally, as in year one, the second year ends with a workshop, where the teachers are prepared for the interventions planned in the final year of the project.

In the subsequent third year, both groups of teachers (T1&T2) are combined to one group (T3), as shown in Fig. 5. The T3 teachers teach their cohort of year three (StY3) with respect to both, the content structure of the electron gas model and context orientation. The students' performance is tested again with pre-post-follow-up-tests. In parallel, the participating teachers' content specific PCK is diagnosed after the last intervention with questionnaires and interviews (see Fig. 8).

The data gained from both teachers and students will be analysed with a mixed methods approach and will be triangulated. The results gained, will then be the basis for a final development cycle of the teaching materials.

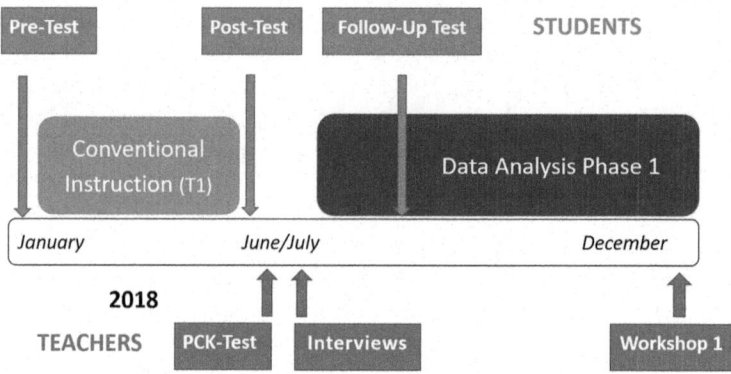

Fig. 6 Research design of year one (2018) of the EPo/EKo project

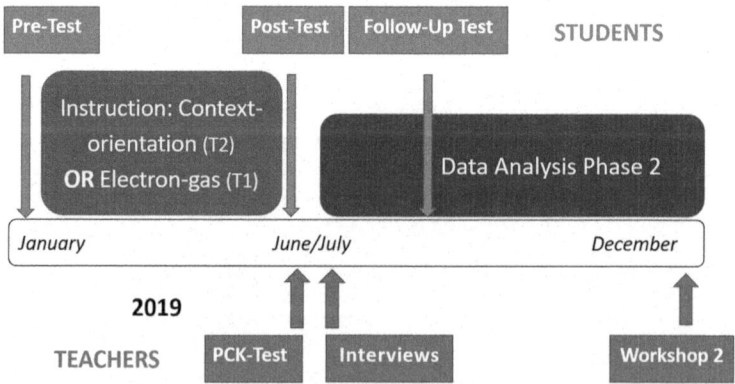

Fig. 7 Research design of year two (2019) of the EPo/EKo project

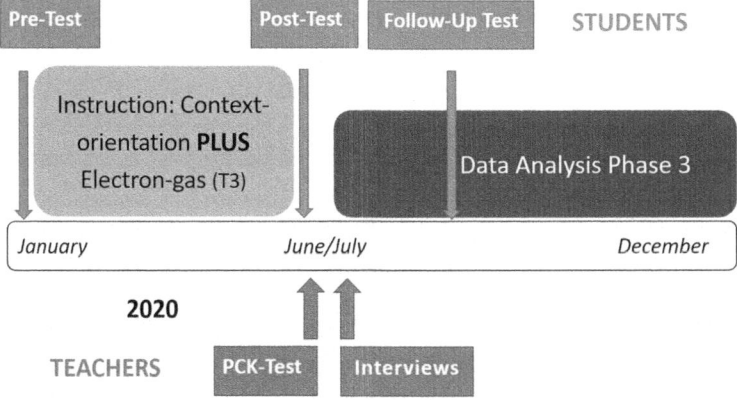

Fig. 8 Research design of year three (2020) of the EPo/EKo project

Current Status of the Project and Outlook

In preparation for the project, a concept test for students' understanding of basic key ideas of introductory electricity has been developed. Existing test instruments [11, 21] have been adapted and expanded for the purpose of our project. Currently, the first intervention phase is taking place with N = 71 classes. In parallel, we are developing a PCK test for the participating teachers. This test will be administered in spring 2018, when the teachers have finished the topic electricity in their classes.

In another strand of the project, teaching materials for the interventions of year two and three are developed in an iterative design process. First of all, materials, which operationalize conventional content structures of introductory electricity with interesting contexts are developed. These context-oriented materials are then implemented in the second year in parallel to a slightly improved version of the already existing content structure based on the electron gas model [3]. Finally, the materials for the intervention of year three are in preparation, which are characterized by the content structure of the electron gas model plus by context orientation.

References

1. McDermott, L.C., van Zee, E.H.: Identifying and addressing student difficulties with electric circuits. In: Duit, R., Jung, W., Rhöneck, C. (eds.) Aspects of Understanding Electricity, pp. 39–48 (1985)
2. Wilhelm, T., Müller, S., Burde, J.-P.: Vergleich von Schülervorstellungen zur Elektrizitätslehre in Hessen und Weißrussland. In: PhyDid B-Didaktik der Physik-Beiträge zur DPG-Frühjahrstagung, Wuppertal (2015). www.phydid.de. Accessed 30 Sep 2017

3. Burde, J.-P., Wilhelm, T.: The electron gas model as an introduction to electricity in middle school science. In: Lavonen, J., Juuti, K., Lampiselkä, J., Uitto, A., Hahl, K. (eds.) Science Education Research: Engaging Learners for a Sustainable Future: Proceedings of ESERA 2015 (Part 1: Strand 1: Learning Science: Conceptual Understanding), pp. 26–36, Helsinki, Finland (2016). ISBN 978-951-51-1541-6
4. Herrmann, F., Schmälzle, P.: Das elektrische Potential im Unterricht der Sekundarstufe I. MNU **37**(8), 476–482 (1984)
5. Schumacher, M., Wiesner, H.: Erprobung des Potentialansatzes in der Elektrizitätslehre in Form einer Akzeptanzbefragungssequenz. In: DPG-Tagung (ed.) Vorträge Physikertagung, pp. 573–578 (1997)
6. Clement, J.J., Steinberg, M.S.: Step-wise evolution of mental models of electric circuits: a "learning-aloud" case study. J. Learn. Sci. **11**(4), 389–452 (2002). https://doi.org/10.1207/S15327809JLS1104_1
7. Gleixner, C.: Einleuchtende Elektrizitätslehre mit Potenzial. Dissertation, LMU München (1998)
8. Schwedes, H., Dudeck, W.G., Seibel, C.: Elektrizitätslehre mit Wassermodellen. Praxis der Naturwissenschaften–Physik **44**(2) (1995)
9. Koller, D.: Einführung in die Elektrizitätslehre. Konzept Lehrer, München. https://www.didaktik.physik.uni-muenchen.de/archiv/inhalt_materialien/einf_elektrizitaet/index.html. Accessed 30 Sep 2017
10. Burde, J.-P., Wilhelm, T.: Concept and empirical evaluation of a new curriculum to teach electricity with a focus on voltage. In: Ding, L., Traxler, A., Cao, Y. (eds.) 2017 Physics Education Research Conference Proceedings, Cincinnati (2017)
11. Rhöneck, C.: Vorstellungen vom elektrischen Stromkreis und zu den Begriffen Strom, Spannung und Widerstand. Naturwissenschaften im Unterricht-Physik **34**(13), 10–14 (1986)
12. Burde, J.-P., Wilhelm, T.: Das Elektronengasmodel im Anfangsunterricht. Praxis der Naturwissenschaften-Physik **65**(8), 18–24 (2016)
13. Burde, J.-P., Wilhelm, T.: Ein Unterrichtskonzept auf Basis des Elektronengasmodells. In: PhyDid B-Didaktik der Physik-Beiträge zur DPG-Frühjahrstagung, Hannover (2016)
14. diSessa, A.A.: Toward an epistemology of physics. Cogn. Instruct. **10**(2–3), 105–225 (1993). https://doi.org/10.1080/07370008.1985.9649008
15. diSessa, A.: A bird's-eye view of the "pieces" vs "coherence" controversy (from the pieces "side of the fence"). In: Vosniadou, S. (ed.) International Handbook of Research on Conceptual Change, 2nd edn. Routledge, New York (2013)
16. Burde, J.-P., Wilhelm, T.: Hilft die Wasserkreislaufanalogie? In: Wilhelm, T. (ed.) Stolpersteine überwinden im Physikunterricht. Anregungen für fachgerech-te Elementarisierungen, pp. 100–104. Aulis Verlag, Seelze (2018)
17. Burde, J.-P.: Konzeption und Evaluation eines Unterrichtskonzepts zu einfachen Stromkreisen auf Basis des Elektronengasmodells. In: Studien zum Physik- und Chemielernen: Vol. 259. Logos Verlag, Berlin (2018)
18. Barab, S., Squire, K.: Design-based research: putting a stake in the ground. J. Learn. Sci. **13**(1), 1–14 (2004)
19. Ejersbo, L.R., Engelhardt, R., Frølunde, L., Hanghøj, T., Magnussen, R., Misfeldt, M.: Balancing product design and theoretical insights. In: Kelly, A.E., Lesh, R.A., Baek, J.Y. (eds.) Handbook of Design Research Methods in Education: Innovations in Science, Technology, Engineering, and Mathematics Learning and Teaching, pp. 149–163. Routledge (2008)
20. Schreiner, C., Sjøberg, S.: The Relevance of Science Education. Sowing the Seed of ROSE. Acta Didactica, Oslo (2004)
21. Urban-Woldron, H., Hopf, M.: Entwicklung eines Testinstruments zum Verständnis in der Elektrizitätslehre. Zeitschrift für Didaktik der Naturwissenschaften **18**(1), 201–227 (2012)

Newton's Apple and Einstein's Time Warp, New Ideas in Teaching Gravity

Warren B. Stannard(iD)

Abstract Over 300 years ago Isaac Newton described gravity as an attractive force between two masses. This concept of gravity is still taught in high schools today. Einstein proclaimed, however, that gravitational effects are the result of a distortion in the shape of space-time. Although this explanation is universally accepted, answering the simple question, 'why do things fall?' is often difficult. This paper introduces graphical and visual analogies and models which are suitable for the introduction of Einstein's General Theory of Relativity at a high school level and provides an answer to the question 'why do things fall?' without resorting to Newton's gravitational force.

Keywords Gravity · Relativity · Time warp · Curved space · General relativity

Introduction

The story is told that a young Isaac Newton whilst sitting under an apple tree observes a falling apple and in a stroke of brilliant insight, creates his theory of gravity [1]. Gravity was, according to Newton, an attractive force that acted at a distance between masses. This theory remains taught in high school classes to this day. Einstein's General Theory of Relativity proposes a completely different description of gravity. The theory proposes that gravitational effects are the result of a distortion in the shape of space-time [2]. This concept is encapsulated in the expression: "Matter tells space-time how to curve, and space-time tells matter how to move" [3].

This idea is often illustrated by the use of an elastic fabric stretched over a frame. A heavy ball is placed in the centre and smaller balls are rolled around the central mass to simulate planetary orbits. The stretched fabric can be interpreted as demonstrating how matter causes spacetime to curve but this can obfuscate the point that it is the time component of space-time that is distorted.

W. B. Stannard (✉)
University of Western Australia, 35 Stirling Hwy, Crawley, WA 6009, Australia
e-mail: w.stannard@uwa.edu.au

© Springer Nature Switzerland AG 2019
E. McLoughlin and P. van Kampen (eds.), *Concepts, Strategies and Models to Enhance Physics Teaching and Learning*,
https://doi.org/10.1007/978-3-030-18137-6_2

13

The question 'why do things fall?' is difficult to answer without resorting to Newton's gravitational force. This paper seeks to provide some simple analogies and models to help students understand gravity in a way which is consistent with Einstein's theory whilst avoiding the mathematical complexity often encountered in advanced studies of General Relativity.

Newton's Apple

Newton observed an apple falling, we are told. His thought process may have followed this line of reasoning: the apple accelerates as it falls—there must be a force acting on it to make it accelerate—as all objects fall at the same rate, this force must be proportional to the objects mass—the force is directed toward the center of the Earth—this is the same force that makes planets orbit the Sun—the force must also be proportional to the mass of the larger object. From this reasoning students can imagine how Newton may have developed his Gravitational Law.

However, we now ask hypothetically, what if Newton had observed an Apple iPhone falling and not an apple? Most good smart-phones have an inbuilt accelerometer and Apps are available to display the accelerometer reading. So, what would Newton have observed? (Students can test this for themselves).

An Accelerometer

An accelerometer measures acceleration. A schematic of an accelerometer is shown in Fig. 1. A small ball is held centrally by two springs. The displacement of the ball from this central position indicates the acceleration of the object to which the accelerometer is attached.

We now ask students, what would the reading on an accelerometer be if it was falling? Many students will say 9.8 m/s^2 but the answer is zero. An object in free fall is weightless; it experiences no force acting on it. The mass and the frame of the accelerometer are falling at the same rate and the mass remains in the center. Students can imagine their weight as it is read by a set of bathroom scales attached to the bottom of their feet whilst they are falling. There would be very little contact between the scales and their feet. The scales would read zero.

Likewise, an accelerometer attached to an object in free-fall would read zero suggesting that there are no forces acting on the object. Now we ask, what does the accelerometer read when it is sitting motionless on the ground? It would appear as displayed on the right in Fig. 2. The 'weight' of the ball would compress the spring and this would correspond to a reading of 9.8 m/s^2 upward although it remains on the ground [4]. On observing this Newton may have come up with a very different theory.

Fig. 1 Three different states for an accelerometer: Top; no acceleration. Middle; acceleration to the right. Bottom; acceleration to the left

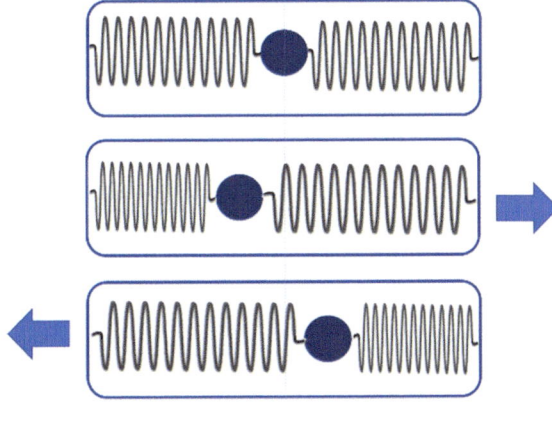

Fig. 2 An accelerometer in free fall (left) and a stationary accelerometer on the surface of the Earth (right)

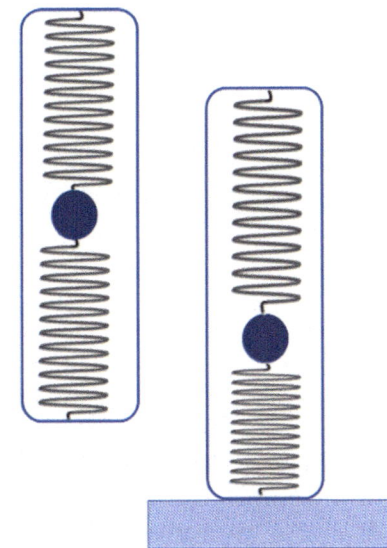

The Space Platform

We now ask students to envisage a space platform and an object floating 'motionless' in space. This object is just floating in space; it experiences no forces and has no acceleration. A platform below this floating object accelerates upwards at a rate of 9.8 m/s^2 until it collides with the floating object. Accelerometers are attached to the object and the platform as shown in Fig. 3. The accelerometer on the space-platform records an acceleration of 9.8 m/s^2, the accelerometer on the floating object reads zero.

Fig. 3 A space platform is accelerated upwards from below a 'motionless' object floating in space. The accelerometers show the respective accelerations

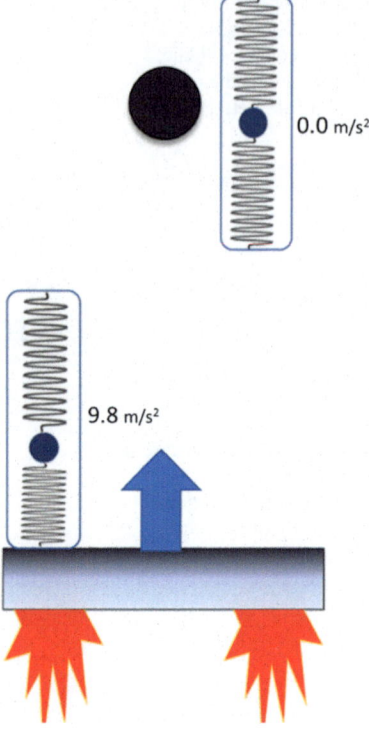

We now consider an object some distance above the Earth's surface falling to ground (see Fig. 4). The space-platform is replaced by the Earth and the diagram is drawn from the perspective of an observer on the Earth.

Comparison of Figs. 3 and 4 suggest that these are two equivalent situations. It could be said that the Earth in Fig. 4 is accelerating upward to meet the 'floating/falling' object just as in Fig. 3, but this would be nonsensical. To explain this equivalence, we need to look at the implications of Einstein's General Theory of Relativity and the meaning of 'curved space'.

Newton's Mysterious Force—A Fable of Two Pilots

The concept of curved space can be illustrated with a fable of two pilots [5]. The pilots, Newton and Einstein, plan to fly from Perth in Western Australia to Durban, South Africa. Using a two-dimensional map, Newton observes that Durban is directly west of Perth. He points his plane due west and sets off on his journey. During his flight he notices that his plane tends toward the North and, to compensate, he banks to the left (or south) to maintain his westerly course. 'I have been pushed north-ward by

Fig. 4 An object falls to ground. This picture is drawn in the Earth's reference frame. The reading on the accelerometers are identical to those in Fig. 3. The two situations are similar

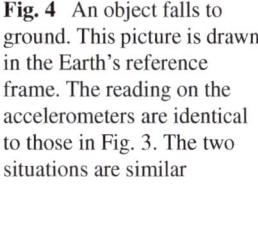

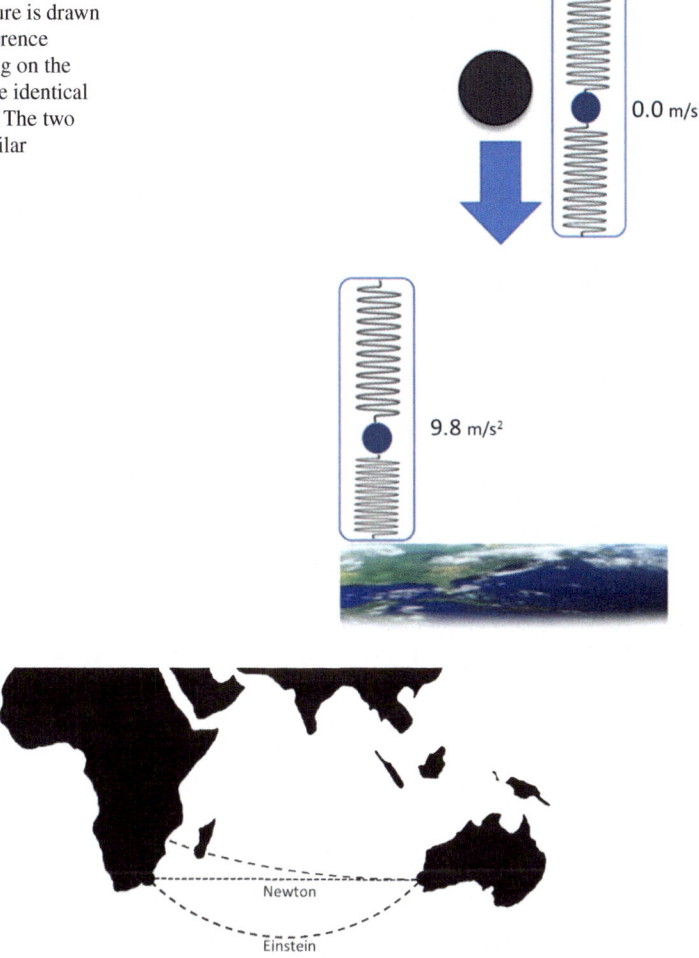

Fig. 5 The two pilots, Newton and Einstein, fly from Perth in Western Australia to their destination in Durban, South Africa, due west of Perth. Newton travels due west but has to bank to the south to maintain his course. Einstein takes the curved path and arrives first

a mysterious force' he proclaims. Einstein, on the other hand, realizes that the map Newton is using is a two-dimensional representation of a three dimensional surface. "We are on a curved surface", he surmises and sets of on a course to the south west as shown. He keeps his plane on an even keel, neither banking to the left or right, and arrives in Durban before Newton. Their flight paths are shown in Fig. 5.

Einstein calculated his flight path by realizing that he was on the spherical surface of the Earth. This path can be derived by drawing scale-lines of equal distance in an East-West direction (see Fig. 6) [6]. The map has lines of longitude running parallel

Fig. 6 Scale-lines are drawn having an East-West separation of 1000 km. By traversing a path perpendicular to each scale-line the shortest path between the two cities can be obtained

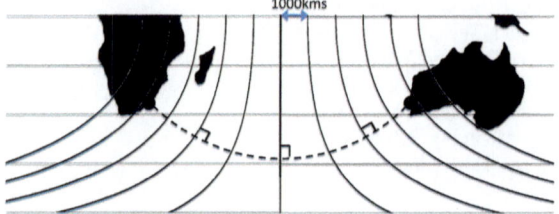

to each other rather than converging at the poles. Distances in an East-West direction at the equator appear shorter on the map compared to the same distance closer to the poles. By first drawing a vertical line joining points equidistant from each city, points at a distance of 1000 km (in an East-West direction) from this central line are joined to create the first curved scale line and so on for each successive scale line. At the equator nine degrees of separation in longitude is approximately 1000 km. Closer to the poles the scale-lines and lines of longitude diverge as shown in Fig. 6.

The shortest path between the two cities is shown by the curved dashed line in Fig. 6. It crosses each scale line at 90°. It is the path that Einstein in our story traversed. It is coincidental with a great circle of the Earth and is the shortest path between Perth and Durban. When Newton pointed his plane due west the natural path of the plane was along the great circle taking him north of Durban. Students can see this more clearly if they use a globe of the Earth. To compensate Newton needed to bank his plane to the left. There was no mysterious force acting on his plane. The 'mysterious force' experienced by Newton is akin to his gravitational 'force'. To Einstein, gravitational effects are the result of the curvature of space-time.

Warped Time

Students are generally familiar with the concept of warped time from popular science-fiction. Einstein's theory predicts that time will proceed at different rates depending on proximity to a large mass. Time warping within the Earth's gravitational influence has been verified experimentally [7, 8] and time warps within a few meters of the Earth's surface can now be measured with the latest atomic clocks [9]. Clocks closer to the vicinity of a large mass will run slower than those far from its gravitational influence.

It is important to first clarify what is meant by time and how is it measured. Time has been defined as "what a clock reads". A clock can be thought of as a device with a periodic oscillating element in which the number of oscillations counted is a measure of time. Atomic clocks are based on the electronic transition frequency within an atom and provide precise measurement of time. It is proposed that all matter has an inbuilt 'clock'.

The term 'clock' will now be used to indicate a particle of unit mass having an inbuilt timing element. The quantity t is a measure of time and is the number of oscillations counted by this clock. 'Relative clock-rate' or just 'clock-rate' will be defined here as the time t recorded on a clock of interest as a fraction of the time t_0 recorded on a reference clock.

$$\text{Clock-rate} = \frac{t}{t_0} \tag{1}$$

To compare two clocks, they need first to be set to zero at an initial time $t_0 = 0$. After a short time, t and t_0 are recorded simultaneously to obtain the clock-rate. The fractional difference between the two times will be called the time warp, $\frac{\Delta t}{t_0}$ where:

$$\frac{\Delta t}{t_0} = \frac{t - t_0}{t_0} = \frac{t}{t_0} - 1 \tag{2}$$

A clock synchronized with a reference clock will have a clock-rate of one. Einstein's theories predict that time will proceed at different rates due to the speed of an object (special relativity) and due to its position in a gravitational potential (general relativity). For example, lifting a clock from the surface of the Earth will result in this clock having a faster clock rate than the reference clock remaining on the ground. Experiments reveal that clocks on the surface of the Earth run slower than those on orbiting satellites. A clock on a GPS satellite at a height of 20,000 km will gain about 45 μs each day relative to a reference clock on the surface of the Earth.

The fractional time warp for a clock at a distance r from the center of a large mass compared to a clock a long way from this mass the can be expressed as follows [5]:

$$\frac{\Delta t_r}{t_0} = -\frac{GM}{c^2 r} \tag{3}$$

If we use a clock on the surface of the Earth as a reference clock then, for heights less than a few hundred meters we can use $g = \frac{GM}{r^2}$ and:

$$\frac{\Delta t_h}{t_0} = \frac{g}{c^2} h \tag{4}$$

Here Δt_h is the time difference at a height h expressed as a fraction of a reference time t_0 measured on the surface of the Earth [6].

Space-Time Plots

Simple Newtonian distance-time plots are drawn on a two-dimensional orthogonal grid with distance on the vertical axis and time on the horizontal axis. A space-time

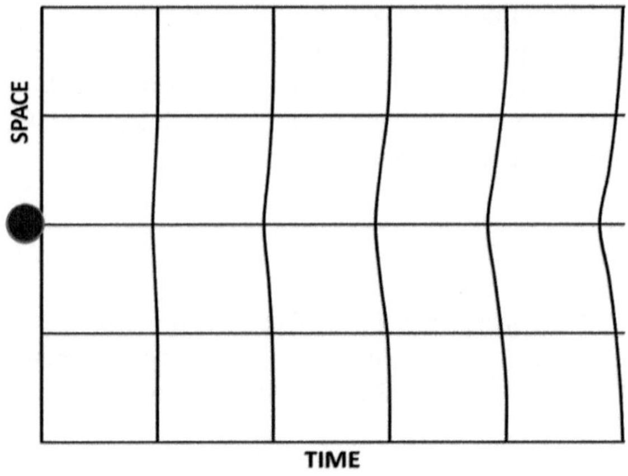

Fig. 7 A large mass such as the Earth distorts space (vertical axis) and time (horizontal axis). It is the time component that is significantly warped

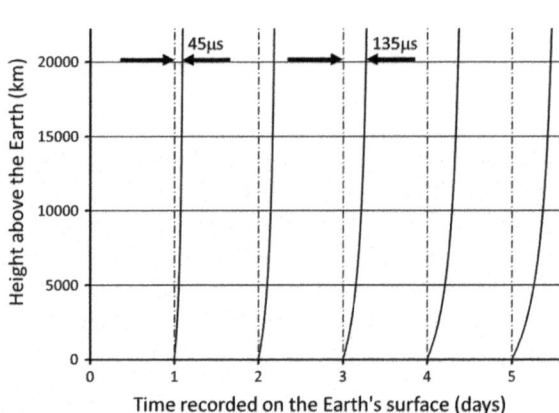

Fig. 8 The effect of time warping in relation to height from the surface of the Earth. The dashed vertical isochrones show time elapsed in days as measured by an observer on the surface if there were no gravitational effects. The solid curved isochrones (not to scale) show the curving of these lines due to the gravitational effects of the Earth

plot usually converts the time units to those of distance by multiplying time by the speed of light. Vertical time lines or isochrones represent a moment in time.

In curved space-time we attempt to draw a curved surface on a flat page and in doing so grid lines become distorted due to the warping of time. This effect is shown in Fig. 7. The distortion is really quite small. Near the Earth's surface a clock runs about 63 μs per day slower than a clock not influenced by gravitational effects.

The effect of gravitational time warping as observed from the viewpoint of the Earth's surface is shown in Fig. 8. To help understand the diagram students can consider two clocks synchronized at time $t = 0$ on the surface of the Earth. One clock is lifted to a height of 20,000 km. After one day the clock at a height of 20,000 km records a greater elapsed time than the clock on the surface by an amount of 45 μs. After two days the time difference would be 90 μs, 3 days 135 μs and so on.

Fig. 9 A flat space-time plot showing the path of a stationary object in space and time

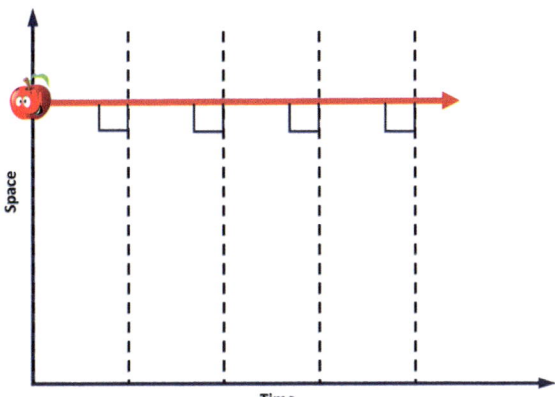

Objects in Free-Fall

Newton's first law states that an object will remain at rest unless acted on by a net external force. A stationary object having no net external force acting on it will follow the path shown by the solid horizontal line in Fig. 9. Time progresses without the object changing its position in space.

According to General Relativity an object in free fall experiences no net force. However, an observer on the Earth will perceive that an object in free fall accelerates downwards towards the surface of the Earth. Newton's second law accounts for the acceleration by the existence of a downward force which Newton called gravity. How do we resolve this conflict?

Inspection of the path of the stationary object in Fig. 9 reveals that when there are no net forces on the object its path on the space-time plot intersects the isochrones at 90°. The object follows the shortest path to the next moment in time. We can call this path a geodesic. The importance of this will become evident when we consider the existence of curved space as shown in Fig. 10. Due to the time warping effects of General Relativity the isochrones on a space-time plot are slightly bent from the vertical. For small heights we can consider g to be constant and the isochrones to be straight lines. In Fig. 10, we see that the vertical isochrones displayed in Fig. 9 are curved due to the proximity of the Earth: 'matter has told space-time how to curve'. The path of an object in free fall intersects the next isochrone at a right angle as shown by the solid red line: 'space-time has told matter how to move'.

For an observer on the surface of the Earth, the object's position changes; it accelerates downward but no force is involved. Figure 11 shows the space-time path for an object falling to Earth from a height of 240 m. The path crosses each curved isochrone at right angles producing a parabolic path in space-time. At this point students can now understand why an object falls; it is not pulled toward the Earth by an attractive force but simply follows a geodesic in space-time.

Fig. 10 A plot of curved (Einsteinian) space-time. The path of an object in free fall intersects the isochrones at right angles as shown

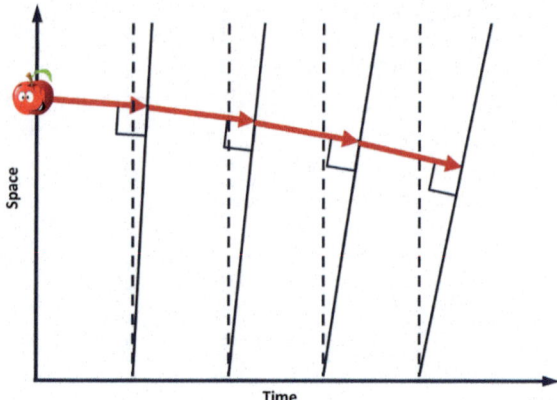

Fig. 11 A free falling object 'falls' toward the surface (shown by the solid red line). After 4 s, an object initially at a height of 240 m above the Earth's surface, has fallen 78.4 m

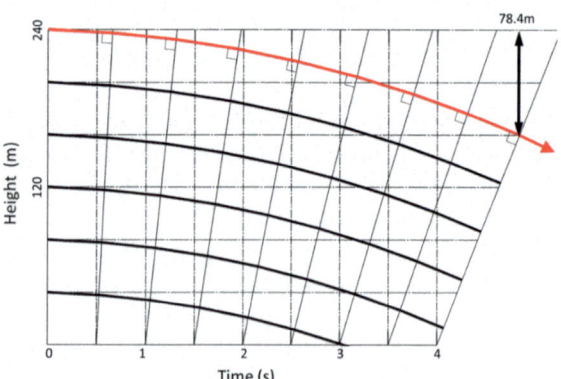

Conclusion

The graphical and visual model presented here is based on two simple principles: that time is warped near a large mass and that objects in free fall follow geodesics in space-time. This model is suitable for the introduction of Einstein's theory of General Relativity at a high school level. It provides an answer to the question 'why do things fall?' without resorting to a gravitational force.

References

1. Stukeley, W.: Memoirs of Sir Isaac Newton's life. http://ttp.royalsociety.org/ttp/ttp.html?id=1807da00-909a-4abf-b9c1-0279a08e4bf2&type=book. Accessed 30 Sep 2017
2. Einstein, A.: Uber die spezielle und die allgemeine Relativitätstheorie, Reprint 2009. Springer, Berlin (1916)

3. Wheeler, J.A., Ford K.A.: Geons, Black Holes, and Quantum Foam: A Life in Physics, p. 153. W. W. Norton & Company (2000)
4. Note: most accelerometers will have the option to be adjusted (or 'calibrated') to ensure that an object on the ground will have an acceleration reading of zero in a vertical direction
5. Gould, R.R.: Why does a ball fall?: A new visualization for Einstein's model of gravity. Am. J. Phys. **84**(5), 396–402 (2016). https://doi.org/10.1119/1.4939927
6. Stannard, W.B., Blair, D., Zadnik, M., Kaue, T.: Why did the apple fall? A new model to explain Einstein's gravity. Eur. J. Phys. **38**(1), 015603 (2017). https://doi.org/10.1088/0143-0807/38/1/015603
7. Vessot, R.F.C., Levine, M.W., Mattison, E.M., Blomberg, E.L., Hoffman, T.E., Nystrom, G.U., Farrel, B.F., Decher, R., Eby, P.B., Baugher, C.R., Watts, J.W., Teuber, D.L., Wills, F.D.: Test of relativistic gravitation with a space-borne hydrogen maser. Phys. Rev. Lett. **45**(26), 2081–2084 (1980). https://doi.org/10.1103/PhysRevLett.45.2081
8. Real-World Relativity: The GPS Navigation System. http://www.astronomy.ohio-state.edu/. Accessed 13 Nov 2015
9. Nicholson, T.L., Campbell, S.L., Hutson, R.B., Marti, G.E., Bloom, B.J., McNally, R.L., Zhang, W., Barrett, M.D., Safronova, M.S., Strouse, G.F., Tew, W.L., Ye, J.: Systematic evaluation of an atomic clock at 2×10^{-18} total uncertainty. Nat. Commun. **6**(6896), 1–8 (2015). https://doi.org/10.1038/ncomms7896

The Virtual Sandbox: An Approach to Introduce Principles of Granular Flow Physics into the Classroom

Erich Reichel⊙, Stefan Radl⊙ and Jakob D. Redlinger-Pohn⊙

Abstract In these lectures students (aged 8–18 years) perform experiments with dry and wet sand. Most importantly, students first encounter contradictory results due to the peculiar nature of flowing granular matter. For example, students measure the density of wet sand as a function of the water content. They also proof the stability of "sandcastles" made of wet sand samples. This ultimately leads to the finding that tiny liquid bridges between the particles must be responsible for this behavior. These findings also support the next step: the students can compare their results with simulations of these experiments. They get an idea of how to simulate such experiments with computer algorithms. The virtual sandbox is the final step used to illustrate the concept of our project: it is a device which shows the combination of real-world experiments and real-time computer simulations. In this way, students can understand the principles behind tools used in modern research and development efforts, specifically in the context of the buzzword "industry 4.0". The feedback of the students indicates significant interest for this topic and documents student gain of knowledge and competences.

Keywords Particle physics · Modelling · Simulation

The Idea—From Experiments to Prediction

From childhood on we are using sand: we play on the beach, build castles, and later houses. But rarely do we worry about why a sandcastle has a stable structure, and

E. Reichel (✉)
University College of Teacher Education Styria, Hasnerplatz 12, 8010 Graz, Austria
e-mail: erich.reichel@phst.at

S. Radl · J. D. Redlinger-Pohn
Graz University of Technology, Inffeldgasse 13, 8010 Graz, Austria
e-mail: radl@tugraz.at

J. D. Redlinger-Pohn
e-mail: redlinger-pohn@tugraz.at

© Springer Nature Switzerland AG 2019
E. McLoughlin and P. van Kampen (eds.), *Concepts, Strategies and Models to Enhance Physics Teaching and Learning*,
https://doi.org/10.1007/978-3-030-18137-6_3

Fig. 1 Disappearing
sandcastle

Fig. 2 Walking on the beach

why it eventually disappears when drying out (see Fig. 1)? Or why do people prefer walking most comfortably along the waterline on the beach (see Fig. 2)? What are the forces that provide sand with certain strength? And how do these forces affect the density of wet sand? Is wet sand heavier or lighter than dry sand?

Objectives of the Project

The presented study deals with the physics of particles and powders, especially sand as a readily available material for performing experiments in the classroom. Along with real-world experiments, numerical simulations of granular matters are introduced. Thereby, the project illustrates two important aspects of basic science:

- how the complementarity of experiments and modelling and simulation studies during the research process can help to advance science, as well as
- how industry may build on this research, e.g., by exploiting modelling and simulation tools for process design, optimization, or even the synthesis of radically new processes.

Specifically, we combined elements of inquiry-based learning with results of modern research in the field of granular matter. This ultimately led to lectures based on an experimental approach, which are adjustable from the primary to the secondary level.

The following objectives have driven the project:

- Introducing a broad motivation for teaching the basics of physics, which helps to develop a general understanding of the importance of technology and engineering.
- Bridging the gap between fundamental knowledge and application to understand the importance of fundamental research.
- Understanding the importance of performing experiments and modelling and simulation studies during the research process.
- Recognizing how modern applications in industry exploit simulations for process design and optimization.

Particle Technology

Processes involving particles and powders in the process industry typically

- are difficult to describe due to the nature of these systems (e.g., cohesive forces affected by water content, electrostatic charging or van der Waals-type forces),
- are hard to engineer due to the lack of sound process models and simulation tools, and
- pose a substantial risk to process stability and safety.

To ensure process and product safety, process simulation, especially the modelling and simulation of particle flows, has been a major trend in the last two decades [1]. This is reflected by a renewed interest of physicists into the field of granular matter, i.e., a sub-class of soft matter. The origins of this development date even back to the seminal work of Bagnold in the 1940s [2]. Particle technology is the corresponding engineering discipline that aims on a technological exploitation of effects observed in granular matter. Accordingly, there is now a strong need to stress the importance of modelling and simulation of particle systems in basic and advanced particle technology courses, as well as in basic physics education.

Therefore, a further aim of this project was to illustrate the actual need to simulate complex processes such as particle systems. This is since many details of such processes are not immediately observable, e.g., due to the opacity of the sand. Also, complex phenomena, such as fluid bridges forming between individual grains, cannot be easily observed. Finally, 3D visuals should support the visualization of complex processes in detail.

The Learning Sequence

The learning setup is provided by Inquiry Based Learning (IBL), whereby the students not only learn fundamentals of physics. Furthermore, they are encountered with the process of achieving scientific knowledge [3]. A 4-step-process will be introduced to the students during the lessons, following the four important phases of a modern scientific approach to research and development:

- *Experiment*—Experiments lead to an observation (i.e., a basic understanding) related to a certain phenomenon.
- *Modelling*—The mathematical description of the experimental data provides a more detailed explanation, and possibly leads to the formation of a suitable model.
- *Simulation*—The model is then translated into a computer code, which allows an efficient variation of the physical parameters. Even parameter sets that are not accessible in the real experiment can be applied, triggering researcher's instinct of exploration.
- *Prediction*—The simulated (calculated) results of the model lead to predictions that may inform researchers and engineers about optimal solutions, or a further need for technological development.

The learning sequences are designed for the primary and secondary level (students aged 8–18 years) with the idea of increasing intellectual sophistication. In higher grades experimental investigations are increasingly used alongside modelling and simulation studies. Using these learning sequences, student's knowledge gain is promoted through the process of scientific inquiry.

Real-World Experiments

In this project, IBL sequences are developed for use across all grades. The core elements of these learning sequences are real-world experiments that lead to questions about phenomena observable in granular matter. These questions are collected in the classroom and lead to experiments and verification of experiments. The experimental results then lead to answers, which help to understand the phenomenon. These results will then be used for a more theoretical approach to investigate certain phenomenon.

All the experiments are suitable for students of the ages from 8 to 18 years. The selection of the experiments and the sequence of them depend on the questions suggested by students. The difficulty level of the particular experiment will be adapted to suit the age of the students. For example, the density experiment described in the next section ranges from simple measuring the mass of different wet sands for the 8-year-old students to calculation of the bulk density of sand for advanced students.

Is Wet Sand Heavier or Lighter Than Dry Sand?—The Bulk Density

If somebody asks to any kind of audience the question above, one will get the answer: "Of course, wet sand is heavier." This answer must be checked carefully. The best way to do this is to measure the density. A certain amount of sand (always the same volume) is weighed on a balance and the density is calculated. Then, the water content of the sand is altered in steps. Care must be taken, that the water sand mixture is stirred until water is homogeneously distributed. A diagram as shown in Fig. 3 will be achieved. Slight changes to this diagram arise from the varying kinds of sand available.

One can notice that the answer is "yes and no". The density depends on the water contents. In certain situations, (i.e., intermediate water content), the density is even lower than that of water (see Fig. 7).

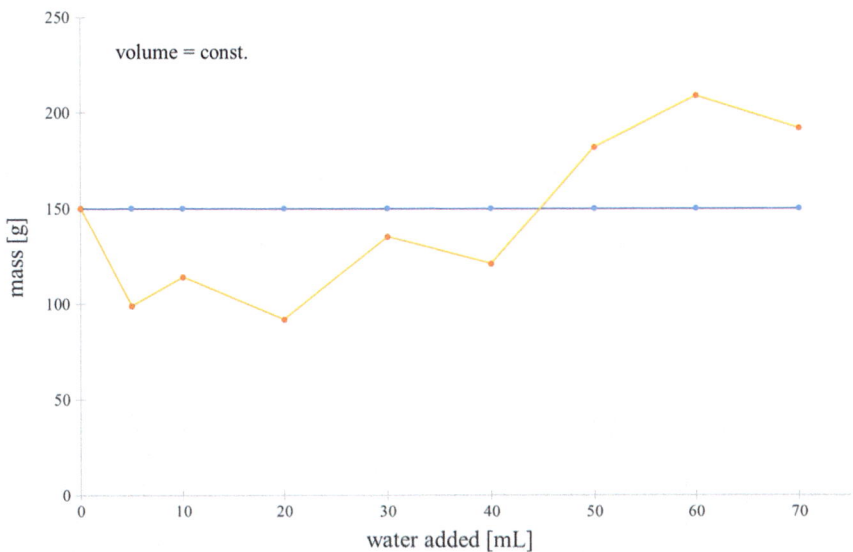

Fig. 3 Mass of samples of dry and wet sand depending on water contents at constant volume (bulk density). The blue line indicates the mass of the dry sand. This measurement was done by 10-year-old pupils

Fig. 4 Stability of "sand cakes" measured with water-filled PET beverage bottles

Stability of Sand Castles

After measuring the density, the "sand cakes" are turned upside down and the stability of the sand-water-mixture is measured. This can be achieved by loading with an increasing amount of weights. These weights can be produced by filling a certain amount of water into PET beverage bottles (Fig. 4). These bottles with increasing weights are placed on top of the "sand cakes" one after each other. For weight distribution a petri-dish is placed between sand cake and bottle. The last weight before a crack is formed gives the maximum load. The results show, that a water content of about 20% lead to the most stable cakes (see Fig. 7).

Reynolds Dilatancy

Figure 5 shows a further amazing experiment with wet sand. If you pressurize sand (i.e., impose a certain amount of deformation), the volume may increase from a certain initial packing density. This phenomenon is called "Reynolds dilatancy." This can be demonstrated with the aid of two beverage bottles, one filled only with water, the other with a mixture of sand and water (care must be taken that sand is densely packed in the bottle). When the two bottles are squeezed, the water in the water-filled bottle rises in the indicator tube. This is not the case in the sand-filled bottle, where the water level drops. This is because of the interstices between the sand grains, which can now hold more water. Putting sand in motion, which is achieved in the present experiment by squeezing the bottle, increases its volume (i.e., the bulk density decreases). The beverage bottle does bulge outwardly not visible to the human eye. Also, an interesting phenomenon during a summer walk on the beach can be explained: "Why does the sand appear dry around my feet when I step on the wet sand?"

Modelling and Simulation of the Experiments

A Brief Description of the Theory Behind the Experiments

The experimental findings allow explanations and the development of models, which ultimately lead to simulations thereof. The examination of wet sand under the microscope shows small liquid bridges between individual grains. These bridges are formed between sand grains and originate from suction in the water bridges (i.e., a local pressure minimum) that is caused by surface tension. This suction ultimately generates cohesive forces between the particles. A good description of this phenomenon can be found in the paper of Schlichting [4], as well as in recent research papers [5].

Figure 6 shows the corresponding ping pong ball model of wet sand. The balls are fixed by tiny water bridges which leads to a stable pyramid. Without water, as in the case of dry sand, no stable pyramid is formed. With an excess of water, the pin pong balls are swimming and no stable pyramid is formed either.

Fig. 5 Illustration of the phenomenon called "Reynolds dilatancy", i.e., the volume expansion of granular material when deformed

Reynolds dilatancy

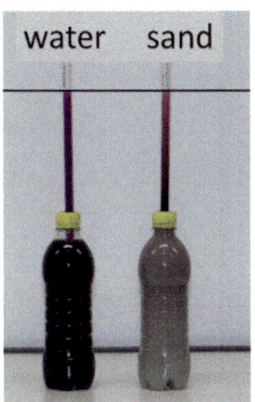
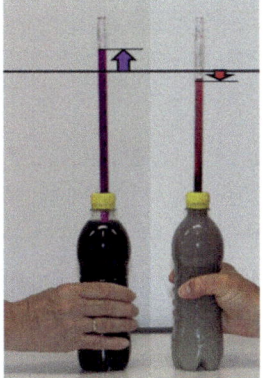

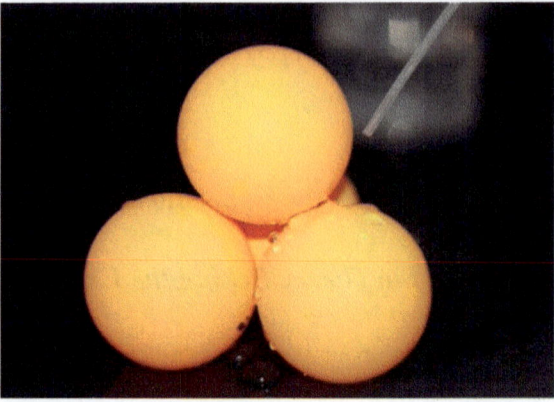

Fig. 6 Ping pong ball model of wet sand

Fig. 7 Bulk density versus water contents for various sand-water mixtures. Inserts illustrate the morphology of samples at some critical water contents from experiments and simulation

Prediction by Means of Simulations

The described model of water bridges enables computer simulations of wet granular matter. Selected results of the numerical simulation can be found in Fig. 7. A more rigorous study how liquid bridge details affect some relevant granular flows, as well as details related to the simulation tool are available in [6]. In addition to the density diagram (see Fig. 3), Fig. 7 illustrates the morphology of sand cakes obtained from the real experiment compared to that of the computer simulation. Both pictures show a convincing match in the results. This shows that the computer code is useful for further investigations.

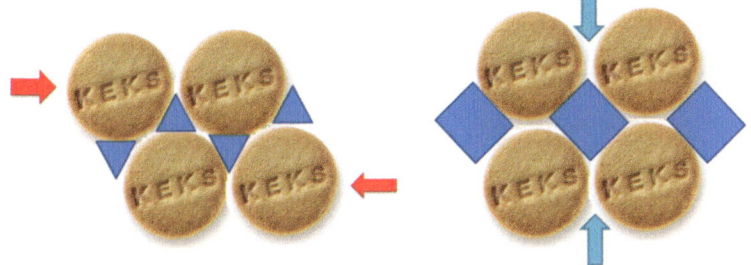

Fig. 8 "Keksperiment" demonstrating Reynolds dilatancy. Red arrows ... movement, blue arrows ... water flow

Reynolds Dilatancy can also be explained quite easily via a simple model—the "Keksperiment" ("Keks" is German for cookie). The cookies take the role of two-dimensional sand grains. The two rows of cookies shown in Fig. 8 can be moved against each other (as indicated by the red arrows). The void space between the cookies increases and a certain amount of water must flow into the larger voids due to continuity reasons. This illustrates the drop of the water level quite well.

Augmented Reality

The Virtual Sandbox project also focuses on augmented reality as suggested by Reed et al. [7]. The Virtual Sandbox consists of a sandpit, a Kinect camera, a beamer and a simulation program. The sand surface is scanned by the camera, and the simulation program calculates the movement of virtual water on the real sand surface in real time. The simulation results are projected onto the sand with the beamer. This allows the loop to be closed from reality to simulation and back to reality (Fig. 9).

The sand surface can be imported into the particle simulator "LIGGGHTS®" [8], and thus the behaviour of the sand can be calculated by the computer, actual sand is "virtualized". This virtualization is an important component of current research topics. Finally, this virtualization is an essential component of "Industry 4.0", i.e., the (predicted) next industrial revolution.

How Do the Learners Perceive the Lectures?

The feedback of 9–10-year-old students is shown in Fig. 10. In general, they liked the topic. For most of them, the question about the weight of dry and wet sand was contradictory. This fact strongly motivated the students to do their research about the phenomenon.

Fig. 9 The virtual sandbox (colors indicate the height levels of the sand; water is animated in dark blue)

During the lessons the students' experimental skills have been documented. Most of the students of all ages were able to perform the experiment with less respectively minimal support. At the end of this lessons the students documented their results. Most of these written documentations show at least a basic understanding of the phenomenon. With increasing age, the water bridge model was used for explaining the experimental results.

Finally, the students understood the phenomenon and they were able to explain it to others. These findings are true not only for the young pupils: even grown-ups were

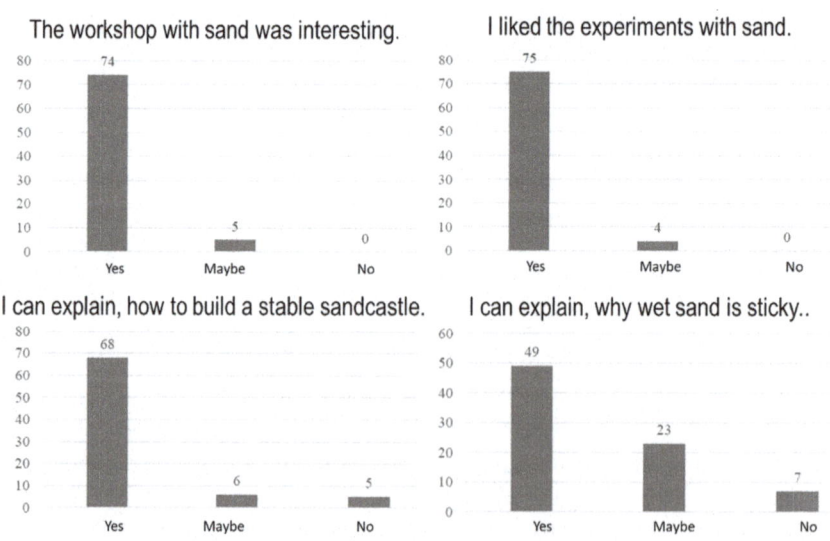

Fig. 10 Feedback of primary level students (Age 9–10), N = 79, female: 43, male: 36

very interested in this topic. Moreover, playing with sand is always stimulating, and it is a prime example to illustrate the way of learning about the nature of science.

Acknowledgements The project was supported by the Austrian Science Fund FWF through project WKP67 (project leader: Univ.-Prof. Johannes G. Khinast).

References

1. Jop, P., Forterre, Y., Pouliquen, O.: A constitutive law for dense granular flows. Nature **441**, 727–730 (2006). https://doi.org/10.1007/s00707-014-1125-1
2. Bagnold, R.A.: The Physics of Blown Sand and Desert Dunes. Methuen, London (1941). https://doi.org/10.1177/030913339401800105
3. Reichel, E., Schittelkopf, E.: Förderung von Kompetenzen durch forschendes Lernen. IMST Newsletter **36**, 4–5 (2011)
4. Schlichting, H.J.: Das Geheimnis der Sandburgen. Spektrum der Wissenschaft, pp. 44–45 (Sept 2014)
5. Wu, M., Khinast, J.G., Radl, S.: A model to predict liquid bridge formation between wet particles based on direct numerical simulations. AIChE J. **62**, 1877–1897 (2016). https://doi.org/10.1002/aic.15184
6. Wu, M., Khinast, J.G., Radl, S.: The effect of liquid bridge model details on the dynamics of wet fluidized beds. AIChE J. **64**, 437–456 (2017). https://doi.org/10.1002/aic.15947
7. Reed, S., Hsi, S., Kreylos, O., Yikilmaz, M.B., Kellogg, L.H., Schladow, S.G., Segale, H., Chan, L.: Augmented reality turns a sandbox into a geoscience lesson. EOS Earth Space Sci. News **97** (2016). https://doi.org/10.1029/2016eo056135
8. Kloss, C., Goniva, C., Hager, A., Amberger, S., Pirker, S.: Models, algorithms and validation for opensource DEM and CFDDEM. Prog. Comput. Fluid Dyn. **12**, 140–152 (2012). https://doi.org/10.1504/PCFD.2012.047457

Three Stories, Three Toolkits: Magnetize Attention, Light up Coloured Ideas, Gas Fantasy into Science

Sara Roberta Barbieri⑩ and Marco Giliberti⑩

Abstract In the context of the European project TEMI (Teaching Enquiry with Mysteries Incorporated), that came to an end in July 2016, the Italian research group of physics education of the University of Milan, developed six experimental toolkits, as legacy materials, about 6 different physics topics (vision of the colours, geometrical optics, electromagnetic induction, electrical circuits, harmonic oscillations, and gas laws). Although the educational paths related to each of them were designed for secondary school, we decided to extend their use to the primary school, selecting two particular topics among those mentioned above (colours and gas behaviour), plus the extra-topic of magnetism, as an extension of the toolkit for the electromagnetic induction. We chose these themes because of their proximity to the children's everyday world, and because they are relevant to their toys (magnets, toy balloons, and colours). Our toolkits, and the lab-sheets included, are based on the 5E's model, that we already used and practiced in the TEMI project. In the present work, children are expected to play freely with the materials contained inside the toolkits and are supposed to do things similar to those performed by their teacher. We will describe the three toolkits and the key concepts that we would like children to face during the lab activities and through storytelling. We will also report on some of the key steps that the stories used to approach particular physics concepts.

Keywords Story-telling · Magnetism · Colours · Pressure · Force field

Introduction

When the European Project TEMI [1] came to an end, the Italian Physics Education Research (PER) group decided to design six different experimental toolkits as legacy

S. R. Barbieri (✉) · M. Giliberti
University of Milan, Via G. Celoria 16, 20162 Milan, Italy
e-mail: sara.barbieri.roberta@gmail.com

M. Giliberti
e-mail: marco.giliberti@unimi.it

© Springer Nature Switzerland AG 2019
E. McLoughlin and P. van Kampen (eds.), *Concepts, Strategies and Models to Enhance Physics Teaching and Learning*,
https://doi.org/10.1007/978-3-030-18137-6_4

37

materials. In fact, during the project, each of the participating countries had to propose a certain number of stimulating scientific "mysteries" that could inspire European teachers that wanted to practice enquiry teaching, with their classes. Therefore, beside the proposals of the mysteries, we felt the need to create something that would help teachers to implement experimental activities and present the mysteries to their students in the classroom. For this reason, we developed six experimental toolkits, each of them based on a particular mystery.

The Experimental Toolkits

To cover different aspects of the secondary school curriculum, relating to physics, we designed toolkit for experimental experiences in:

- States of matter, with particular attention to gases and gas laws;
- Geometrical optics, with particular attention to the concept of refraction;
- Vision of the colours, with particular attention to the addition of light and fluorescence;
- Mechanical oscillations, with particular attention to harmonic oscillations;
- Electrical circuits, with particular attention to circuits in series and in parallel, and the internal resistance of a battery;
- Electromagnetic induction, with particular attention to the phenomena related to eddy currents.

These toolkits were then trialled in continuing professional development workshops with secondary school teachers and with groups of secondary school students participating in Scientific Degree Plans (whose acronym in Italian is PLS). In both of these contexts we implemented the enquiry method together with the use of the experimental toolkit and we always observed increased motivation, that led to teachers becoming more creative and designing new experiments to be carried out with their students [2, 3].

The results obtained with these secondary school teachers and students, encouraged our group to extend the project to use with younger students and in particular with primary school students. The work we present here pertains to this student group. Of course, while maintaining the essential spirit of the project TEMI, to engage student's attention and motivation by means of a mystery, (to which, a sort of submersion in physics must follow, to get the solution of the mystery) we had to radically change the approach in order to involve younger students. Primary school students can be involved through observing fascinating phenomena, they can also enjoy wow-effect experiments, but to evoke their scientific reasoning, to stimulate their ability to question and inquire into the world around, children need more.

The Story-Telling

In fact, as reported in the literature [4], and as we already observed with our previous activities in primary schools [5, 6], a very fruitful way to involve pupils in science is through story-telling. In our experience, in fact, story-telling has at least a double purpose: on the one hand, it can focus children's attention as soon as the story begins, and on the second hand, the story constitutes a sort of canvas along which all the related scientific activities can be carried out in a coherent manner, that is the story alternates with experiments within an enquiry approach to physics.

As Corni et al. report [4], story-telling is a way of involving students that is important for two main reasons: the formation of *hypothesis* by means of figurative thinking (that is elicited by the use of fantasy, metaphors, rhythm and emotionality) and the *motivation*, or the stimulus in *making questions* (by means of imaginative and romantic thinking).

In our approach we have taken these two reasons and we have used stories as a thread that runs through all classroom activities, so as to inspire and guide students, and the storytelling is interrupted a number of times so as to introduce experiments. In one case we also used the storytelling for another reason. We realized that a story can have a powerful and alternative way to describe and/or introduce an *abstract physical entity* (like pressure, in our case), which may otherwise have been difficult to be dealt with directly and by definition would have been too challenging for children and connected to a much too sophisticated concept.

Enquiry activities are fundamental, especially for children, in order to stimulate their curiosity and enable them to develop new concepts. Guided by the teacher, students will describe what they are doing and what they are expecting to find out. Therefore, we designed three educational paths for children, that brought together story-telling and laboratory experiments. We started from topics that were part of pre-existing toolkits, we selected three of these based on their appeal to children's lives. We adapted the toolkit contents in each case, we pointed out the key concepts of each of the three experiences and finally we wrote the stories centered on these concepts. Among the topics treated in the TEMI project, we selected three of them, avoiding mathematical tools as much as possible, or descriptions intrinsically too quantitative.

So the topics selected were:

- The vision of colours, with the mystery toolkit "Guess the colour!";
- The magnetic phenomena and the introduction of the concept of force field in physics, using part of the mystery toolkit "The invisible brake";
- The introduction of the concept of pressure and the observation of the behaviour of gas, with the mystery toolkit "Help! The bottle is eating the toy balloon!".

The next sections will describe in more detail the key concepts that we wanted to deal with, the parts of the stories where the concepts are treated and the laboratory activities we chose that would allow pupils to become familiar with the concepts introduced by the story.

Story and Toolkit 1: Magnetic Field Used to Introduce Force Fields in Physics

The physical concept of field is very complicated, even for secondary school students, who usually consider it has the same properties as a force. For the electrostatic field or for the gravitational field, this temptation appears justified considering how they are defined (force/charge or force/mass). But, what about magnetic field or even a static magnetic field? This entity rotates and attracts or repels a magnet, a simple force is not sufficient to describe the magnetic field and, besides, what is the "magnetic charge" that should be used to determine the magnetic force experienced by a magnet?

The aim of our story is to present the magnetic field, not in a formal way, but in a playful way, through a description of its actions on a magnet and through its phenomenology in the world around us.

The Story: "King Neodimio's Hug"

The protagonist is Mariblù, a playful young-girl that, during the narration, children will discover to be a magnet. Her uncle is King Neodimio, who knows how magnets work and slowly leads Mariblù to understand who she is, and the reason for her blue hair and her red body. In the final part of the story Mariblù realizes that the peculiar tickling that she feels when her uncle Neodimio is close to her is due to the many lines that he continuously lets out and recovers.

Nobody can feel this tickling, except a magnet—in other words this means that nobody can feel a magnetic field except a magnet or something that has magnetic properties. The co-protagonist is a real girl who starts playing with Mariblù.

At the beginning the two girls ignore the existence of their magnetic properties. But then they start to understand each other and together they discover the magnetic properties of some of the objects around them. They explore the world they come across: the ferromagnetic properties of nails, keys and so on. And, finally they encounter King Neodimio, Mariblù's uncle. He enters the scene and has the strongest force, and most importantly he has the strong personality of a neodymium magnet!

Through understanding the effect of the king, the girls discover the magnetic field. The story ends when Mariblù and her friends play together forming a circular dance, as Fig. 1 shows. So, looking carefully, children will recognize that the orientation of the friends is not arbitrary. That is, they all obey a rule, and children can gradually try to understand this rule, and decide if this drawing is a correct image of a field (as represented in Fig. 2).

Proceeding in this way, it is then possible to position only the source magnet and every other magnet will know exactly how to orientate around the source magnet. The fields lines of the source magnet act as a map, or a guide, for all the other magnets around it, as shown in Fig. 3.

Fig. 1 The round dance that all the Mariblù's friends plays. Children can notice that the red part of each magnet is attached to the blue part of another magnet

Fig. 2 The field lines of the source magnet come from the disposition of the magnets that can feel its presence, because of their magnetic nature

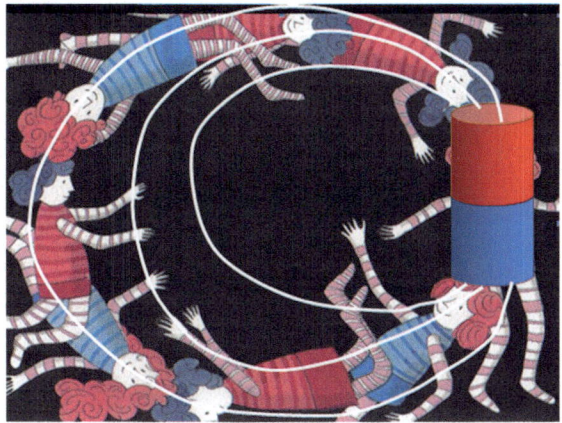

Fig. 3 The field lines of the source magnet are a sort of guide for the disposition of every other magnet placed near the source

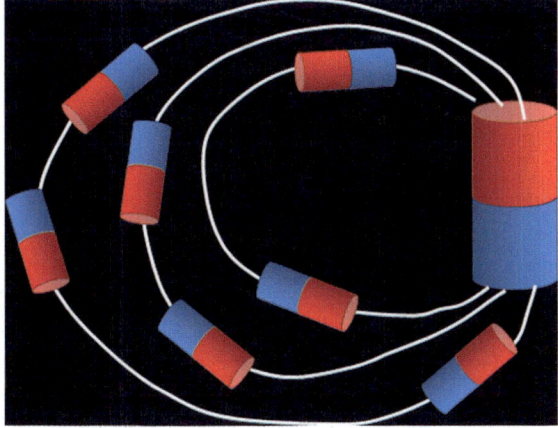

This approach sketches a magnetic field around a source magnet and only if you are a magnet you can see it, and like a map, it tells you how to rotate at the point you are at.

In general, hence, a *field* is a drawing around a source: you can see it only if you are made of the "same stuff" as the source. As a map, the field will tell you how to behave. It should be clarified to children that Mariblù's blue hair and her red dress could be changed with other colours without changing the story, and the same could be done for the round dance and all the other things, as well.

The Experimental Toolkit 1: Discovering the Magnetic Field

This toolkit is used to draw magnetic fields using very simple materials, such as, thin lead, the inner part of adhesive tape, cardboard and some magnets. The source magnets are attached to the cardboard using the adhesive tape, while a cylindrical magnet (that can be called an explorer magnet) should be put inside the plastic inner part of the adhesive tape, this can be considered as a container for the explorer magnet. The system of the explorer magnet plus it's container can be considered as a kind of rough magnetic probe for the magnetic field generated by the magnets attached to the cardboard.

Now, with the help of a teacher, children can move the probe around the cardboard and they will notice that the explorer magnet rotates if it is displaced in the magnetic field. In this way, it is possible to draw the direction of the magnetic field line point by point, as shown in Fig. 4.

Story and Toolkit 2: Vision of Colours

The focus of this second topic is mainly the fascination of children that listen to the story in the dark: the story will be interrupted to leave room for activities with

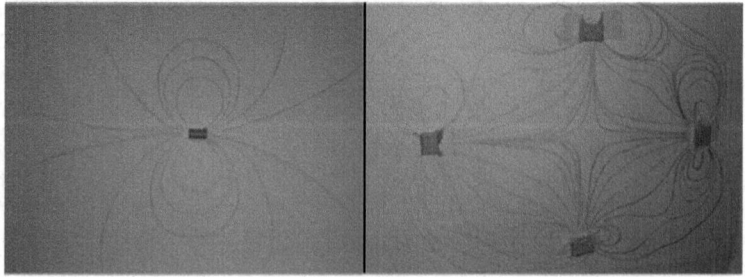

Fig. 4 Magnetic field lines drawn by a 10 years old child: a dipolar field and a quadrupolar field

coloured lights and to appreciate the effects that light can have. All the concepts involved will be treated qualitatively, but not superficially, especially in relation to the vision of colours. In fact, at the end of the learning path children should be able to answer the question of: "Why is this object red and the other object is green or blue?".

Another objective of this educational path is to create familiarity with fluorescence, a phenomenon that is widely present even in children's world, although it is not always associated with this term or recognized as such.

The Story: "In the Darkness"

The story tells about Milla, a young girl that is raised free and brave. One day something pushes her into the "endless darkness" that had always frightened the inhabitants of her village. A frog is her "sherpa in a mountain made of darkness". Milla lives two parallel adventures: one in the world pertaining to the fascinating phenomena of coloured lights and the colours in general; and a second world inside herself and through the experience of darkness she discovers things that otherwise she would never have found.

The story is not included here because it is quite long and it should be used to introduce children to the fairy world as a sort of theatrical experience, where the scene is dominated by coloured lights that are highlighted by the darkness and by the suspense of the storytelling.

The Experimental Toolkit 2: Vision of the Colours

The topics involved in the story and experimented directly by the children using the toolkit "Guess the colour!" are [7]:

- Observation of *monochromatic lights*; children can use lasers of different colours (red, green and blue) on gratings, a hair, slits, CD or a little sieve.
- Observation of *non-monochromatic* lights; children can use a transmission diffraction grating with LED lights (red, green, blue, white), that in some cases can be considered quite monochromatic; and then incandescence bulbs, PC screens, spectral lamps (not included in the toolkit) and at the end children can give meaning to a rainbow produced by the light passing through a prism.
- Observation of a number of *coloured cards* under monochromatic lights; children will see the coloured cards change their colours depending on the colour of the monochromatic light used to illuminate them, as it is shown in Fig. 5. Along with the observation it is useful to have the children fill in a table in order to recognize patterns between the colour of the light used to illuminate the cards and the colour the cards appear to be.

Fig. 5 Images of a set of coloured cards under red light (on the left) and under white light (on the right)

Fig. 6 Comparison between a white (on the left) and a yellow (on the right) card illuminated using a blue laser

- Observation of a yellow card under blue light. In this case, something strange seems to happen: the card appears to be so much brighter than all the other coloured cards; while the other cards reflect a fraction of the light that have struck them, the yellow card under blue light seems to generate light instead of appearing almost dark. Moreover the colour of the light emitted is completely different from the colour of the monochromatic light used to strike the card. This new phenomenon that amazes the observer is called *fluorescence*. Together with the observations on cards, the toolkit also contains other materials that produce fluorescence under blue light (e.g. Plexiglas, tonic water, Silly Putty) and materials that produce fluorescence under green light (Coca Cola, tea, olive oil). Figure 6 compares the illumination of a white card (completely reflective) and a yellow card, that is fluorescent under blue light.
- Observation of the overlap of two monochromatic lights; using a black cardboard with an hole and two monochromatic LED lights at the same time, it is possible to appreciate the *additive synthesis* of lights (an example in Fig. 7), from which it is possible to deduce, qualitatively, some considerations about colour mixing when painting - this phenomenon is called subtractive synthesis.

Fig. 7 Additive synthesis of light: magenta comes from the overlap of blue and red lights

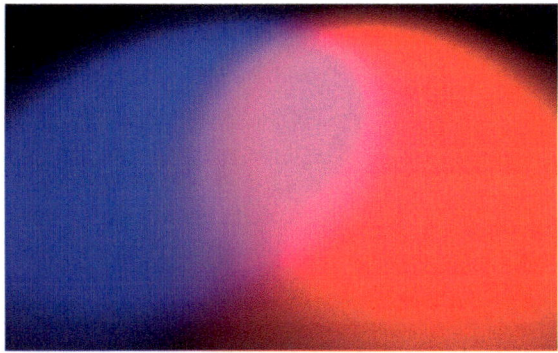

Story and Toolkit 3: The Concept of Pressure and the Gas Laws

In this last case, the story is used to describe and suggest to children the concept of pressure, and it is not used to connect to laboratory activities. The aim of this story is to introduce the problematic concept of pressure, whole avoiding formal or mathematical tools, in a context where the concepts of volume and temperature have been already introduced. The toolkit "Help! The toy bottle is eating the toy balloon!" is used in the second part of the activity which allow children to learn how the behaviour of a gas depends on three variables: volume, temperature, pressure. This story is thought to have sufficient details so as to allow students to carry out subsequent experiments involving the concept of pressure.

Pressure is not an intuitive entity. It is more abstract then what one would initially consider. Even the word, pressure, is widely present in everyday life, and therefore a risk is that the term is often confused with something that has to do with force. Instead, pressure is (in the simplest cases) a scalar and an isotropic quantity and it is possible to build an idea of it as a consequence of a mental model.

We found this definition to be too formal and not appropriate to introduce the concept of pressure to children. We therefore decided to remain in the domain of what is dreamy and imaginative. For this reason, a story seemed to be a good way to proceed.

The Story: "the Strange Ears of Grandpa Gay-Lussac"

This story is told through the voice of a boy and using his own easy language (although this boy is no less than Gay-Lussac's grandson). The boy's kinship with his illustrious grandfather is revealed at a certain point of the story, when the physical ideas of pressure start becoming important.

"Is a caress a thing?". It is this type of question that starts a discussion about abstract concepts. The discussion is kept very close to the emotional and daily world of children, so that, for example, an alley cat is involved...

"There is almost no difference between a real caress, a caress in a dream, or a caress in my mind!"—these are preparatory steps to get the fact that before you measure something, you must have imagined it or dreamt it, or otherwise you will have no chance to measure it, because "reality" is never so trivial.

Gradually the concept of pressure emerges, until the boy says:

"Stop being mysterious. Things are more or less like that: even if there is no breeze, or the windows are closed... in any place where we can breathe, there..." and continues: *"...right there, there is a gas, which brings with itself what my grandfather described to the world: pressure!"*.

Narration continues, but there is another leap to make. The unification of gas and liquids in a unique category called fluids. In fact, the behavior of fluids can be described using the concept of pressure. The water of the sea is practically the same as the air in the atmosphere, because they are both fluids attracted by the center of the Earth. And it is at this point, the concept of hydrostatic pressure is introduced: that pressure increases as the fluid depth increases—in the story we refer to "the strange ears of Gay-Lussac". That is, Gay-Lussac's ears have the amazing property of being able to hear the laughs of pressure and gas when they travel around the world on their exciting journeys: regardless of the orientation of the ears, Gay-Lussac always hears the same quantity of laughs if the depth of the fluid is the same.

The Experimental Toolkit 4: Gas Laws and the Hydrostatic Pressure

This toolkit aims to explore in different ways the relations between two of the physical quantities of volume, temperature and pressure. In particular, the:

- Relationship between *volume and temperature*. There are at least two interesting experiments. The first is carried out using a bottle containing some water and sealed with a toy balloon. The system is then heated up and the toy balloon gradually increases its volume at the top of the bottle. The second experiment is a sort of a reverse of the previous. In this case the bottle contains hot water at its bottom, then a toy balloon is used to seal the top of the bottle and the system is left for a few minutes at ambient room temperature. As the temperature of the water inside the bottle decreases, the volume of the gas in the system diminishes and an amazing effect will appear. The effect is that of a bottle eating a toy balloon. Another typical experiment that can be realized very easily is that of a can containing some water that has to be heated and suddenly submerged (with the help of a tongs) in a large bowl containing cold water. The can will instantly collapse and a typical pop sound will be heard.

- Relationship between *pressure and volume*. It is possible to visualize the relationship between the volume and the pressure of a gas contained in a system by means of simple experiments that can be carried out with materials contained in the toolkit. In many cases these experiments can be carried out with common kitchen items. In the first basic experiment, a plugged syringe containing air is positioned vertically using a clamp. The piston of the syringe has to be in such a way that it should be possible to add a number of equal weights on its top. After the addition of each weight, the volume of the air contained in the syringe will be reduced and can be easily measured. The use of qualitative or quantitative reasoning will allow for a description of the relationship between the pressure and volume of the gas. Another experiment that could be useful to examine the relationship between volume and pressure in a gas is the "Cartesian devil". It is also useful to clarify to students that pressure is not a vector and emphasize that pressure is *isotropic*. The same relationship between pressure and volume can be visualized in a very effective way while observing the shape of air bubbles moving from the bottom of a cylinder filled with glycerol. This last experiment can be considered in conjunction with the final topic, that is hydrostatic pressure.
- *Hydrostatic pressure*. Using a bottle filled with water and with holes located at various different heights from the bottom, it is possible to observe the effect that the pressure of the liquid increases as the hole approaches the bottom. There are many other simple experiments that can also show the effects of atmospheric hydrostatic pressure, e.g. using a plastic cup and a postcard or a beaker completely submerged in a large bowl containing coloured water. All of these ideas are well known experiments (suitable even for elementary school students) but we have mentioned them here in order to give an idea of the educational path that we have in mind and that can be carried out using the third toolkit.

Conclusions

Storytelling is a very important tool in children education because it stimulates the formation of hypothesis, generates motivation and gives children the opportunity to pose questions to themselves, and this is because children are free to use their own imaginative and romantic thinking.

A teacher can use stories that already exist and are written by others, but the story-telling can have even greater stimulating and engaging effects when the teachers develop their own stories. In this case, teachers become more aware of the key concepts involved and are more enthusiastic in their retelling of the story. This creativity can lead to increased motivation for learning physics for both the student and the teacher.

The origins of this study is based on practical experience with secondary school students and reviewing literature about the importance of story-telling in involving young students and introducing them to physical concepts. This current research was based mainly on how to write a good and engaging story, rather than on the most common difficulties children have with particular physics topics. In fact, in 2011 we already conducted a study with primary school children where we wrote another story, "Mommy Comet", to deal with the birth of the life in our planet, the motion of objects thrown on the surface of the Earth and some basic elements of relativity. That experience gave us a level of sensitivity to calibrate the depth of treatment of the topics involved in this work - how to stay close to physics, the use of correct definitions and how to use imagination, fantasy and a fairy atmosphere of a story. We are strongly aware that children's interest is captured in the darkness, with the images projected and the story-telling.

On the other hand, the experiments that were performed while reading the story, make us believe that the story is not detached from the physics that we want to transmit, even if children have not yet developed a deep knowledge of the physical concepts. However, developing a familiarity with science, the sense of discovery of something, the idea that certain things are connected and not others, the fact that connections have to be widely discussed to be understood, and that an appropriate discussion can generate science knowledge, these are experiences we would like to give to young children.

References

1. TEMI. European Community's Seventh Framework Programme (FP7/2007–2013) under grant agreement n° 321403 - 2012-1. http://www.projecttemi.eu/ last accessed 2017/09/30
2. Barbieri S.R. et al.: Teachers participant to the European Project TEMI practice the enquiry methodology in their classroom. In: Proceedings of The International GIREP-EPEC Conference 2016, July 6–10: Key competences in physics teaching and learning. P.102, Institute of Experimental Physics, University of Wrocław, Pl. M. Borna 9, 50-204 Wrocław, Poland, ISBN: 978-83-913497-1-7 (2016)
3. Barbieri S.R. et al.: Enquiry for Physics Teachers Following the TEMI Methodology. In: Proceedings of The International GIREP-EPEC Conference 2015, July 6–10: Key competences in physics teaching and learning. p. 52, Institute of Experimental Physics, University of Wrocław, Pl. M. Borna 9, 50-204 Wrocław, Poland, ISBN: 978-83-913497-1-7 (2016)
4. http://www.univ-reims.fr/site/evenement/girep-icpe-mptl-2010-reims-international-conference/gallery_files/site/1/90/4401/22908/29321/29596.pdf, last accessed 2017/09/30
5. Barbieri S.R. et al.: "Mommy comet". Short stories for introductory mechanics. Problemy Wczesnej Edukacji. 3(15) pp. 133–136 (2011)
6. Barbieri S.R. et al: "Mommy comet" brings children to discover the Solar System. In A. Lindell, A.-L. Kähkönen, & J. Viiri (Eds.), Physics Alive. Proceedings of the GIREP-EPEC 2011 Conference. Jyväskylä: University of Jyväskylä (2012)
7. Barbieri S.R. et al.: "Guess the colour!" a mystery to approach the vision of the colour. Chemistry in action! 107, p. 42 (2016)

Mathematics—A "Quasi-natural Science" at School?

Eduard Krause (ID)

Abstract "The most distinctive characteristic which differentiates mathematics from the various branches of empirical science, and which accounts for its fame as the queen of the sciences, is no doubt the peculiar certainty and necessity of its results" (Hempel in Am Math Mon 52, [1]). In this way, Hempel described a specific feature of modern mathematics. Hempel traced this certainty and this necessity back to the consideration that after Hilbert mathematics has developed towards a formal-abstract science, which does not need ontological bounding to reality (physical space) anymore but is based solely on logical consistency. However, at school mathematics still is not and cannot be taught in this abstract form for various reasons, here the usage of illustrative material, diagrams and real-world application is essential. From a constructivist point of view, students in this environment construct a so-called "empirical-concrete" mathematical worldview on the basis of real objects than a "formal-abstract" worldview as it is described by Hempel. This "empirical-concrete" mathematical worldview, on an epistemologically level, is quite close to the natural sciences; one could speak of mathematics in modern school as something like a "quasi-natural science". Questions regarding core beliefs, terms and processes of natural sciences (education) in general and physics (education) in particular are of eminent importance to mathematics teachers from this point of view and open a whole range of possibilities for further discussions and cooperation at the crossroads. Research- and teaching-projects of the University of Siegen already aim to use this synergies. On the level of interdisciplinary research on knowledge processes, historical research, research regarding epistemic beliefs, textbook analysis, design based research etc. we try to make these synergies between the participating educational disciplines fruitful.

Keywords Interdisciplinary · Relation between mathematics and physics · Epistemology

E. Krause (✉)
Universität Siegen, Herrengarten 3, 57072 Siegen, Germany
e-mail: krause@mathematik.uni-siegen.de

© Springer Nature Switzerland AG 2019
E. McLoughlin and P. van Kampen (eds.), *Concepts, Strategies and Models to Enhance Physics Teaching and Learning*,
https://doi.org/10.1007/978-3-030-18137-6_5

Why Should Physics Educators and Math Educators Work Together?

A Subsection Sample

"God has practiced geometry by creating the world!" [2]—metaphysical convictions like this have prompted scientists like Johannes Kepler to explain the real world through the connection of natural science and mathematics. Thus, in a Heureka-experience, he could explain the apparently irregular distances of the planets through his polyhedron model using platonic bodies (Fig. 1). Even without this metaphysical accessory, the connection between mathematics and physics can be reconstructed not only from the historical genesis of both subjects, it also offers numerous possibilities for teaching and learning these disciplines. Especially in higher education, teaching often crosses the subject-borders. Interdisciplinary has been an increasing trend since the beginning of the 2000s, not only at universities as a new mode of research [3, p. 21] and teaching [4] also as a didactic principle in the design of teaching-learning processes in the secondary school, for example in form of problem-oriented projects [5, 6].

In the hope that students will develop key competencies, Granzer [7, p. 75ff] also speaks of "cross-curricular competencies", the teaching practice varied variants of this demanding and complex learning concept (see, for example, [8, 9]). However, the

Fig. 1 Keplers polyhedron model [2] as an example how mathematics and physics can be thought together

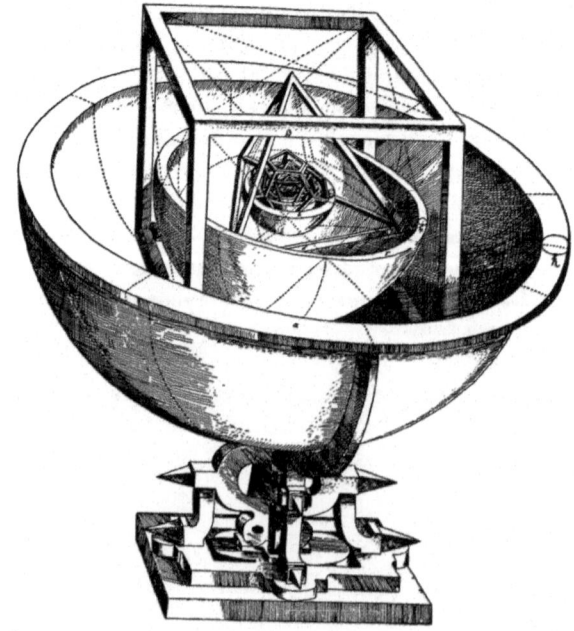

focus is always on the ability to systematically combine, apply and reflect knowledge from different disciplines. The didactic potential of subject-linking teaching and learning has been discussed for some time, especially STEM subjects [10–13].

The subjects of mathematics and physics are particularly suitable for this purpose [14]. A view in the mathematics core curriculum immediately reveals that an important task of mathematics teaching is to enable pupils to experience mathematics in application situations with real material contexts [15, 16]. Although there are already a number of different approaches to the embedding of non-mathematical application contexts into the teaching, unfortunately these are partly still revealed as standard calculation techniques [17, 18] which are covered in exceptional mathematical situations, or a large number of different application cases are opened can be executed consistently and can be confusing [c.f. 19]. Physics offers many contexts to apply mathematics in an authentic way, but the use of these contextualizations in the process of teaching and learning mathematics is not as easy as it seems.

On the other hand, mathematics plays an important role in learning physics [20–22]. You can't do physics in a quantitative way without mathematics. Mathematics is used as a tool for calculations and as a language to articulate complicated content in an exact and short way. It is a big challenge for the physics teacher to make clear why and how mathematics is so important for physics. But these correlations of mathematics and physics are on a content level. This paper discusses the connection of math and physics on a meta-level, to show the epistemological parallels between school mathematics and physics and what they imply for teaching and learning mathematics and physics. These parallels on an epistemological level and the connection on a content level make a solid base to teach mathematics and physics in an interdisciplinary way in school, in the teacher training and to design research projects.

Beliefs About Modern Mathematics and Mathematics in School

First, we have to think about what mathematics generally is. When you ask people what mathematics is, you would get very different answers. Someone who learned modern Mathematics on a university level would agree with Einstein [23] when he says (quoted from [1]): "*As far as the laws of mathematics refer to reality, they are not certain; and as far as they are certain, they do not refer to reality.*" The certainty of mathematics, which means internal consistency, is very important for mathematics. But when you refer mathematics to reality, you lose this certainty. This abstract-formal (formalistic) notion of mathematics was developed by Hilbert [24] about 100 years ago, and led to a tremendous increase in productivity in mathematics, as the subject area of mathematics was considerably expanded [25]. For example, in the geometry only spaces up to dimension three were investigated before Hilbert; because spaces of a higher dimension were regarded as non-existent with regard to our physical space [26 p. 171f]. Since Hilbert [24] Mathematics is seen as the archetype of formal-abstract sciences and can be characterized by these points:

- axiomatic structure
- detached from reality
- a science of uninterpreted abstract systems (focus on structures)
- the absolute notion of truth (internal consistency)

Summarized you can say with Freudenthal [27, translated by the author]: *"The bond to reality has been cut by Hilbert"*.

According to a well-documented hypothesis in mathematics didactics the belief of mathematics, which students acquire in the classroom, is different [28]. Attempts to mediate mathematics in an abstract-formal sense at school have failed in the 1960s and 1970s ('New Mathematics'). Instead, teachers try to convey mathematics vividly and with the help of materials and media. Empirical studies reveal the following dilemma: While some teachers would like to convey an abstract-formal mathematical theory, their pupils actually learn an empirical-representational theory about illustrative means [29, 30]. Students and teachers then have different conceptions of mathematics—an observation which is fundamental for a further development of mathematics teaching. The theories of the pupils are then 'physical' theories in the sense that phenomena of reality, namely, the visualization, are described and explained. The common feature, however, is their empirical character: it is about describing and explaining empirical (physical) phenomena.

Here are two examples: The first example is school geometry. The Euclidean geometry was first fully axiomatized by Hilbert [24]. The axiom system has 20 axioms. In the geometry teaching of all school forms, this abstract-formal axiom system is not put to the beginning, but the geometry is "illustrated" with the help of figures, which are constructed on drawing sheets or on the blackboard. However , students do not acquire the abstract-formal geometry in the meaning of Hilbert's Foundation of Geometry, which has only been skillfully illustrated with the aid of drawing-figures, but an empirical-representational theory of drawing-figures [28]. This is confirmed by empirical investigations by Schoenfeld [31, 32] who has observed students in solving geometric problems and whose "belief system", their view of geometry has been investigated. It assigns an "empirical belief system" to the learner, as opposed to the "mathematical belief system" of mathematicians.

The second example is school analysis. Objects of modern differential and integral calculus are functions. These are algebraic objects or abstract-formal relations which satisfy certain theoretical properties. Since the field of the real numbers is treated in school only in an approach, and in particular, the completeness, as a rule, is not adequately addressed, calculus in school cannot be taught in the sense of the university. The visualization of functions by means of curves, that is to say by sections of their drawn graphs, instead form a basis for the lesson and are often regarded by the pupils as the actual objects of instruction [30]. According to the thesis, the pupils acquire an empirical theory about curves, which are based on drawing sheets or on the display of graphing calculator.

However, an empirical-objective understanding of mathematics in school doesn't need to be a deficit. Historical analyzes [33] show that many mathematicians had such an idea in the history of mathematics: examples are Moritz Pasch for geometry and Leibniz as well as Jakob and Johann Bernoulli in calculus. These historical case studies suggest that an empirical-physical conception of mathematics seems appropriate in particular for school teaching.

An empirical belief-system of mathematics can be characterized by the following points [31]:

- A real subject area
- Explanation of Knowledge by logical-conclusively leading back to known knowledge
- Secure knowledge through empiricism.

Epistemological Parallels Between School Mathematics and Physics

Since Galileo Galilei, not only the induction but also the deduction plays an important role in physics [34]. Inductive generalization of natural phenomena alone does not lead to scientific knowledge. Although there have been attempts to establish induction as a systematic method for the formation of theory in physics, these reflections have not led to success. The most famous of these types is the inductive method of Francis Bacon, who attempted to gain new insight (a line result of the first stage, second stage, etc.) by systematic observation and notation (the well-known bacon plus minus tables) in an iterative process. Although he has discovered a connection between heat and movement with this method, Simonyi comments in his standard work on the history of the physics [35, p. 215, translated by the author], "*In practice, Bacon's detailed method of finding true laws of nature cannot be used and no new law has been discovered in this way.*" Because this misconception of the nature of physics is so widespread, Einstein was up against a purely inductive understanding of physics in numerous writings on the philosophy of natural science. Holton comments on Einstein as follows [36, p. 379, translated by the author]:

> The ascension of a logical step-ladder, in which induction of general propositions from an existing set of individual observations is concluded by means of induction, corresponds to a science in the youthful state.

Einstein himself said [23, p. 84, translated by the author]

> There is no inductive method that leads to the basic assumption of physics. The fact that this has not been recognized is the fundamental philosophical error that so many researchers of the nineteenth century have killed.

Empirical Sciences describes Einstein as follows [23, p. 105, translated by the author]

> Science is an attempt to reconcile the chaotic multiplicity of sensory experiences with a logically unified system of thought; in this system the individual experiences with the theoretical structure have to be related in such a way that the resulting assignment is unique

and convincing. The sensory experiences are the given object. But the theory which it is intended to interpret is human work. It is the result of an extremely tedious adaptation process: hypothetical, never quite final, always subject to questions and doubts.

In a letter, he wrote to his friend Maurice Solovine in 1952 he sketches his understanding of the nature of natural science, which is known in the literature as an EJASE model (reprinted and commented on by Holton [36], Kuhn [37] or Krause [38]. The core of his explanation is the sketch in Fig. 2.

Line E represents the experiences. But the real starting-point of the process, according to Einstein, is the axioms, which can be seen as a free, mental creation, and thus man-made. The curved arrow to the axioms, which is called a "jump", certainly raises the most questions in this model. Einstein comments on him literally (Einstein to Solovine, quoted from Holton [36, p. 378f, translated by the author]):

> Psychologically, the A is based on E. There is no logical path from E to A, but only an intuitive (psychological) connection, which is always on the point of revocation.

What exactly is involved in the creation of the axioms of a theory is difficult to say—certainly empirical experiences, but also other "bold ideas", which may also have arisen from metaphysical background convictions. From the set of axioms and definitions, sentences can be derived (S in Fig. 2). The connections of A and S are pure deductions, which in themselves are not subject to any empirical influence. The deduced propositions must prove themselves in empirical theory on empiricism. This makes Einstein clear through the arrows from S to the straight line E. In physics, this is usually done through experiments. The axioms A, the deduced propositions S, and their connection form the mathematical nucleus of the theory. The concepts of the theory have in themselves no empirical reference. In this model, there is another addition to the Galileo-themed concluding deduction and induction-abduction. The creation of the axioms is not an inductive generalization of empirical sensory experiences. The arc from E to A is more likely to be identified as an abduction. This logic was introduced by the logician Charles Sanders Peirce (1839–1914) into the knowledge theory around the turn of the 19th century to the 20th century [39]. The starting

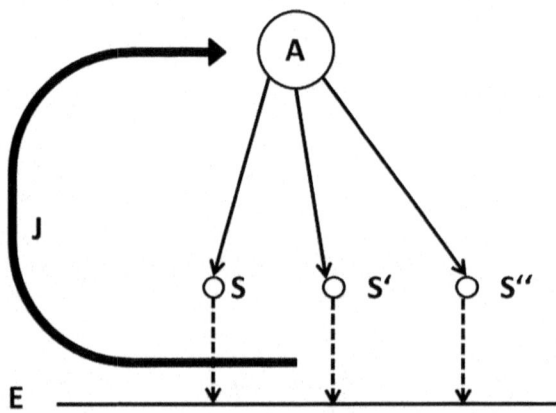

Fig. 2 Einsteins EJASE-modell

point of abduction is usually a surprising fact, a question or a still inconsistency, which suggests a rule that answers the question or removes the doubt. This rule is not deduced from the assumptions but is set ad hoc. From these axioms, sentences are deduced in a further logical conclusion. Empirical data obtained empirically are used to test these theorems. Thus, abduction, deduction, and, in a certain sense, induction also play a role in this process of knowledge generation. Of course, this process will not be strictly chronological. The formation of the theory is improved by repeating these steps.

The EJASE modell Einstein as an exemplary model to describe the nature of science has parallels with the empirical belief of mathematics, which Schoenfeld [31] describes. Table 1 shows an overview over the points of comparison.

To illustrate the parallels between mathematics in school and physics, we want to have a closer look at the task, which Schoenfeld [31] used for his research:

> Two intersecting straight lines and a point P on one of the two lines are given. Show how a circle can be constructed using only a compass and a ruler in that way that the circle owns the two straight lines as tangents and P as a contact point (Fig. 3).

The transcripts[1] made Schoenfeld [31] construct the archetype of students behaving like "pure empiricists" generating and verifying their Ideas and assumptions exclusively by drawing. A mathematician, in contrast, uses terms and definitions. The students' review of a geometric hypothesis is solely done based on a drawing. Accuracy will increase its' usefulness while the mathematician does not bother to draw his solution. Figure 4 shows that logical reasoning is not as important to students

Table 1 The points of comparison between school mathematics and Einsteins EJASE modell

EJASE modell	School mathematics
E	A real subject area
A → S	Explanation of knowledge by logical-conclusively leading back to known knowledge
S → E	Secure knowledge through empiricism

Fig. 3 A geometric task to find out the belief-system

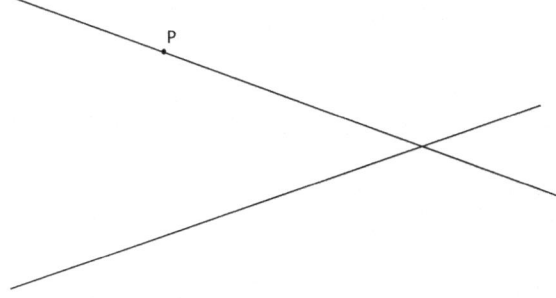

---

[1]Schoenfeld generates his category-system by analyzing transcripts of students and not using surveys like various other researches with a view on beliefs.

Fig. 4 Students accept both
solutions even though they
are logically inconsistent,
because the picture looks
appropriate [31, p. 170]

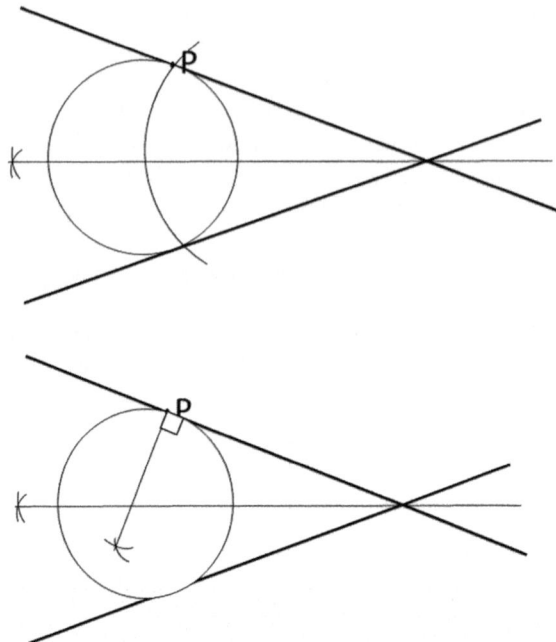

as a right-looking drawing because they are able to accept two different (logically
inconsistent) solutions if both look appropriate.

Summary and Potential for Research

Physics and mathematics have a large content intersection. Beside these similarities there are epistemological parallels that allow comparing these two disciplines.
Especially mathematics in school that differs from modern mathematics in its ontological bounding to reality shows interesting points of comparison with physics. In
this sense school mathematics can be described as quasi-natural science. This comparison includes research-potential for both disciplines: In foundational research the
exact synergies should be explicated and what they imply for teaching and learning.
This research should be accompanied by empirical research that e.g. identifies the
beliefs of mathematics and NOS of students against this theoretical background.
Further some interventions should be designed and evaluated. For example, an interdisciplinary course in the teacher training, where math and physics teacher discuss
the similarities and differences of this two subjects on a content level and on an
epistemological meta-level.

References

1. Hempel, C.G.: Am. Math. Mon. **52** (1945). Reprinted in: Feigl, H., Sellars, W. (eds.) Readings in Philosophical Analysis. Appleton-Century-Crofts, New York (1949). Reprinted in: Newman, J.R. (ed.) The World of Mathematics, vol. III. Simon and Shuster, New York (1956)
2. Kepler, J.: Harmonice mundi (1619). Reprinted in: Caspar, M. "Gesammelte Werke", Band 1, Mysterium Cosmographicum, 405 (1995)
3. Defila, R., et al.: "Interdisziplinarität" in der wissenschaftlichen Diskussion und Konsequenzen für die Lehrerbildung. In: Wellensiek, A. (ed.) Beltz Wissenschaft: Bd. 38. Interdisziplinäres Lehren und Lernen in der Lehrerbildung. Perspektiven für innovative Ausbildungskonzepte 17. Beltz, Weinheim, Basel (2002)
4. Schier, C., et al. (eds.): Pädagogik, Interdisziplinarität und Transdisziplinarität als Herausforderung akademischer Bildung: Innovative Konzepte für die Lehre an Hochschulen und Universitäten. De Gruyter, Bielefeld, Berlin (2014)
5. Labudde, P.: Fächerübergreifender naturwissenschaftlicher Unterricht – Mythen, Definitionen, Fakten. Zeitschrift für Didaktik der Naturwissenschaften **20**(1), 11 (2014). https://doi.org/10.1007/s40573-014-0001-9
6. Moegling, K.: Kompetenzaufbau im fächerübergreifenden Unterricht: Förderung vernetzten Denkens und komplexen Handelns; didaktische Grundlagen, Modelle und Unterrichtsbeispiele für die Sekundarstufen I und II, vol. 2. Prolog-Verl, Immenhausen bei Kassel (2010)
7. Granzer, D.: Interdisziplinäres Lehren und Lernen - Leitgedanken für einen neuen Lernbereich. In: Wellensiek, A. (ed.) Beltz Wissenschaft: Bd. 38. Interdisziplinäres Lehren und Lernen in der Lehrerbildung. Perspektiven für innovative Ausbildungskonzepte 72. Beltz, Weinheim, Basel (2002)
8. Caviola, H.: Wie Fächer miteinander ins Gespräch kommen: Modelle der Fächervernetzung und ihre Lernziele. In: Caviola, H., et al. (Hrsg.) TriOS: Nr. 2/2012. Interdisziplinarität, pp. 5–36. LIT-Verlag, Münster (2012)
9. Labudde, P. (ed.): Naturwissenschaften vernetzen - Horizonte erweitern: Fächerübergreifender Unterricht konkret (1. Auflage). Klett/Kallmeyer, Seelze-Velber (2008)
10. Kelly, R.T., Knowles, J.G.: A conceptual framework for integrated STEM education. Int J STEM Educ **3**, 11 (2016). https://doi.org/10.1186/s40594-016-0046-z
11. English, L.D.: Advancing elementary and middle school STEM education. Int J Sci Math Educ **15**, 5–24 (2017). https://doi.org/10.1007/s11858-008-0106-z
12. LaForce, M., Noble, E., King, H., Century, J., Blackwell, C., Holt, S., Ibrahim, A., Loo, S.: The eight essential elements of inclusive STEM high schools. Int J STEM Educ **3**, 21 (2016). https://doi.org/10.1186/s40594-016-0054-z
13. Michelsen, C.: Expanding context and domain: a cross-curricular activity in mathematics and physics. ZDM **30**, 100–106 (1998)
14. Galili, I.: Physics and mathematics as interwoven disciplines in science education. Sci. Educ. **27**, 7–37 (2018). https://doi.org/10.1007/s11191-018-9958-y
15. Winter, H.: Mathematikunterricht und Allgemeinbildung. In: Mitteilungen der Gesellschaft für Didaktik der Mathematik **61**, 37 (1996)
16. Büchter, A., et al.: Schulmathematik und Realität-Verstehen durch Anwenden. In: Bruder, R., Hefendehl-Hebeker, L., Schmidt-Thieme, B., Weigand, H.-G. (Hrsg.) Handbuch der Mathematikdidaktik, pp. 19–50. Springer, Heidelberg (2015). https://doi.org/10.1007/978-3-642-35119-8_2
17. Jahnke, T.: Zur Authentizität von Mathematikaufgaben. In: Beiträge zum Mathematikunterricht, 38 (2005). https://doi.org/10.17877/de290r-5732
18. Baumann, A.: Eine kritische Betrachtung zum Thema "Modellierungsaufgaben" anhand von Beispielen aus dem hessischen Zentralabitur 2009. Mathematikinformation **55**, 15–23 (2011)
19. Witzke, I.: Zur Problematik der empirisch gegenständlichen Analysis des Mathematikunterrichtes. Der Mathematikunterricht, 2/60, 19 (2014)
20. Krey, O.: Zur Rolle der Mathematik in der Physik. Wissenschaftstheoretische Aspekte und Vorstellungen Physiklernender. Dissertation an der Universität Potsdam (2012)

21. Uhden, O.: Mathematisches Denken im Physikunterricht – Theorieentwicklung und Problemanalyse. Dissertation an der Technischen Universität Dresden (2012)
22. Pospiech, G., et al.: Mathematik im Physikunterricht. Themenheft in: Naturwissenschaften im Unterricht – Physik **153/154**(27) (2016)
23. Einstein, A.: Aus meinen späten Jahren. Deutsche Verlagsanstalt, Stuttgart (1952)
24. Hilbert, D.: Grundlagen der Mathematik. Leipzig (1900)
25. Davis, P.J., Hersh, R., Marchisotto, E.A.: The Mathematical Experience, Study Edition, Boston (1995)
26. Möbius, F.: Der barycentrische Calcul. Barth, Leipzig (1827)
27. Freudenthal, H.: Die Grundlagen der Geometrie um die Wende des 19. Jh., Vortrag auf der DMV Tagung in Utrecht (1972)
28. Struve, H.: Grundlagen einer Geometriedidaktik. BI, Mannheim et al. (1990)
29. Schlicht, S.: Zur Entwicklung des Mengen- und Zahlenbegriffs. Springer, Heidelberg (2016)
30. Witzke, I., et al.: Domain-specific beliefs of school calculus. J für Math Didaktik **37**(1), 131 (2016). https://doi.org/10.1007/s13138-016-0106-4
31. Schoenfeld, A.H.: Mathematical Problem Solving. Academic Press, Orlando (1985)
32. Schoenfeld, A.H.: How We Think. A Theory of Goal-Oriented Decision Making and Its Educational Applications. Routledge, New York (2011)
33. Witzke, I.: Die Entwicklung des Leibnizschen Calculus. Eine Fallstudie zur Theorieentwicklung in der Mathematik. Franzbecker, Hildesheim (2009)
34. Militschenko, I., et al.: Entwicklungslinien der Mathematisierung der Physik – die Rolle der Deduktion in der experimentellen Methode. Der Mathematikunterricht 5/63, 21–29. Friedrich Verlag, Seelze (2017)
35. Schimony, K.: Kulturgeschichte der Physik. Verlag Harri Deutsch, Thun, Frankfurt (1990)
36. Holton, G.J.: Thematische Analyse der Wissenschaft: Die Physik Einsteins und seiner Zeit (1. Aufl.). Suhrkamp-Taschenbuch Wissenschaft, Teil 293. Frankfurt am Main: Suhrkamp. Retrieved from http://d-nb.info/810165260/04 (1981)
37. Kuhn, W.: Das Wechselspiel von Theorie und Experiment im physikalischen Erkenntnisprozess. In: Praxis der Naturwissenschaften – Physik, 12, 355–362 (1983)
38. Krause, E.: Das EJASE Modell Einstein als Ausgangspunkt physikdidaktischer Forschungsfragen – Anregungen aus einem Modell zur Natur der Naturwissenschaft. Physik und Didaktik in Schule und Hochschule (PhyDidA), 1/16, 57 (2017)
39. Peirce, C.S.: Collected Papers of Charles Sanders Peirce. Harvard University Press, Cambridge (1985)

Part II
Innovation in Undergraduate Physics Education

A 5E-Based Learning Workshop on Various Aspects of the Hall Effect

Dominique Persano Adorno⊙, Leonardo Bellomonte and Nicola Pizzolato⊙

Abstract Learning activities in constructivist environments are characterized by active engagement, inquiry, problem solving, and collaboration with peers. The 5E learning cycle is a student-centered instructional model for constructivism, where the students perform five phases of instruction: Engagement, Exploration, Explanation, Elaboration, Evaluation. The purpose of this contribution is to present a 5E-based learning path of advanced physics aimed at strengthening Physics/Engineering student understanding about the quantum Hall effect, a phenomenon observed at low temperatures in a two-dimensional electron gas subject to a strong perpendicular magnetic field. The quantum Hall effect, a rare example of microscopic effects observable on a macroscopic scale, allows us to establish very precise values of microscopic quantities, such as the electron charge and the Planck constant. In the present learning path, we stimulate a discussion about the integer and fractional quantum Hall effects, aimed at introducing a unified picture based upon composite fermions, interacting quasiparticles that may be viewed as fermions carrying attached a fictitious magnetic flux. Finally, we discuss the quantum effect in graphene, the 'miracle material' for its unique and exceptional properties.

Keywords 5E learning cycle · Hall effect · Graphene

D. Persano Adorno (✉) · L. Bellomonte
Department of Physics and Chemistry "Emilio Segrè", University of Palermo, 90128 Palermo, Italy
e-mail: dominique.persanoadorno@unipa.it

L. Bellomonte
e-mail: leonardo.bellomonte@unipa.it

N. Pizzolato
IISS "Pio La Torre", Via Nina Siciliana 22, 90135 Palermo, Italy
e-mail: nicola_pizzolato@libero.it

© Springer Nature Switzerland AG 2019
E. McLoughlin and P. van Kampen (eds.), *Concepts, Strategies and Models to Enhance Physics Teaching and Learning*,
https://doi.org/10.1007/978-3-030-18137-6_6

Introduction

In this contribution we present and discuss a 5E-based learning path of advanced physics aimed at strengthening Physics/Engineering master level student understanding of the quantum Hall effect, a phenomenon observed at low temperatures in the nearly free electron gas of a layered material (2D) subject to a longitudinal electric field and to a strong perpendicular magnetic field. The characterizing features of the Hall effect and the associated quantum transport phenomena in 2D systems have been the focus of much attention in the last decades, but the quantum Hall effect explanation is very often beyond the scope of the usual undergraduate courses and outside the experience of most non-specialist physicists [1, 2]. However, a deeper understanding of the peculiarities of the electron behavior in a layered material is essential in undergraduate education of students of electronic engineering, physics, and material science.

An effort is here devoted to fill this gap. The Hall effect is peculiar of systems with free or nearly free charged particles whose motion is affected by a magnetic field in presence of an electric field and represents the most remarkable manifestation of Landau quantization. Moreover, it is a rare example of microscopic effects observable on a macroscopic scale, allowing us to establish very precise values of microscopic quantities, such as the electron charge and the Planck constant. In the present learning path, we stimulate a discussion about the integer and fractional quantum Hall effects, aimed at introducing a unified picture based upon composite fermions, interacting quasiparticles that may be viewed as fermions carrying attached a fictitious magnetic field [3, 4].

We have proposed a learning path to deepen physics and engineering undergraduates' understanding of the fundamental concepts underlying the electronic properties of new 2D materials, graphene in particular [5]. The workshop was based on the introduction and application of basic concepts of Solid State Physics, rarely used in learning paths, such as the symmetry properties of the crystal lattice and a simple application of group theory, features of electron energy bands, and the degeneracy of wave functions [6]. In our pilot-study the workshop has been greatly appreciated by the students, mainly for the following reasons: (1) challenge to deal with graphene; (2) opportunity to act as researchers; (3) favorable occasion to practically use basic concepts of quantum mechanics. In the wake of these results, for this learning activity we have also chosen a constructivist environment, characterized by active engagement, inquiry, problem solving, and collaboration with peers. In this context, the teacher, rather than a dispenser of knowledge, is a guide, a facilitator and a co-explorer who encourages the learners to question, challenge, formulate their own ideas and opinions, and draw conclusions [7]. In this way knowledge is acquired through involvement with content, instead of imitation or repetition [8].

The 5E learning cycle is a student-centered instructional model for constructivism [9], where the students perform five phases of instruction: *Engagement* (students activate and assess prior knowledge by connecting the "new to the known"), *Exploration* (students explore a real-world problem), *Explanation* (students explain their thinking), *Elaboration/Extension* (students elaborate on their reasoning and solidify their understanding), *Evaluation* (students should assess their own learning). The 5E cycle is a pedagogical approach with the purpose of developing students' critical thinking and helping them to explore and evaluate their own learning.

Our intent is that at the end of the 5E-based workshop students have consolidated their knowledge about the Hall effect and achieved a more meaningful understanding of relevant concepts regarding the quantum behavior of correlated particles. In particular, they should be able to make an analogy between the Cooper pairs in superconductor materials and the composite electrons in graphene, and to appreciate analogies and differences between the concepts of effective mass of electrons in semiconductors and effective charge of Dirac fermions in graphene.

The Five Phases of the Workshop

Engagement

The Engagement phase involves setting up the learning environment in a way that stimulates student interest and generates curiosity about the topic under study. It gets students personally involved in the lesson, while assessing prior understanding. During the Engagement stage, students first encounter and identify the instructional task, then make connections between past and present learning experiences, and organize ground work for upcoming activities. The discussion should stimulate students' curiosity and encourage them to ask their own questions. We here start recalling the classical Hall Effect and its use in experiments to derive the sign, density and mobility of the charge carriers [10].

The Hall effect, discovered by Edwin Hall in 1879, is the setup of a voltage difference across a metallic thin film, subject to an electric ($\mathbf{E}_x$) and a magnetic field ($\mathbf{B}_z$). Consider a 2D conductor lying in the xy plane, and apply both fields. The electrons will start to accumulate on one y-side of the bar, depleting the opposite side. Consequently, an electric field $\mathbf{E}_y$ is set up and, at equilibrium, its strength is such as to impede further charges to accumulate. The total force $\mathbf{F}$ (Lorentz force) acting on a charge $-e$ moving with velocity $\mathbf{v}$ is, in *SI* units

$$\mathbf{F} = -e(\mathbf{E}_x + \mathbf{v} \times \mathbf{B}_z). \tag{1}$$

The two contributions in Eq. (1) influence the electrons lying in the Fermi sphere by impressing on them both a longitudinal and a centripetal acceleration. The effect of the random collisions, with average time τ, acts as a frictional damping term in the equation of motion:

$$\mathbf{F} = d\mathbf{p}(t)/dt + \mathbf{p}/(\tau) = m^*[(d\mathbf{v}(t))/dt + \mathbf{v}/\tau], \tag{2}$$

where m^* is the electron effective mass. At equilibrium $(d\mathbf{p}(t)/dt = 0)$, the current I is constant. The induced electric field $\mathbf{E}_y$, called the Hall field, is given by

$$\mathbf{E}_y = \mathbf{E}_x \times e\mathbf{B}_z\tau/m^*. \tag{3}$$

Its value, deducible from the above equations subject to the condition $\mathbf{v} = const$, is set up by the electron polarization along y. It prevents further charge accumulation and balances the Lorentz force (see Fig. 1). This effect is called *Hall effect*.

Some other quantities or effects deducible from the above relations, valid at equilibrium, will be of central interest in the following discussion [10]. They are:

1. the *cyclotron frequency*, $\omega_c = e\mathbf{B}_z/m^*$. It is the frequency of the rotatory motion impressed in the xy-plane by the Lorentz force acting on a moving free charge (one or more complete rotations are only possible at low-temperatures);
2. the *Hall coefficient*, $R_H = E_y/j_xB_z = -1/ne$. It is proportional to the Hall field, depends on the sign and value of the surface charge concentration, and is often used to determine both quantities;
3. the material resistivity or *magneto-resistance* ρ_{xx}. It depends on carrier density and mobility and measures the electric resistance in presence of a magnetic field orthogonal to the electric field;
4. the *Hall resistivity* ρ_{xy}, defined as the ratio V_H/I where V_H is the Hall voltage along the y-side. It is proportional to B_z.

The classical Hall effect is the precursor of a cascade of "modified" Hall effects, some of them peculiar to 2D materials, that have been discovered successively due to advancements in quantum mechanics and technology: (i) the Integer Quantum Hall Effect, (IQHE) which occurs at low temperature in presence of strong transverse magnetic fields; and (ii) the Fractional Quantum Hall Effect (FQHE), which is a

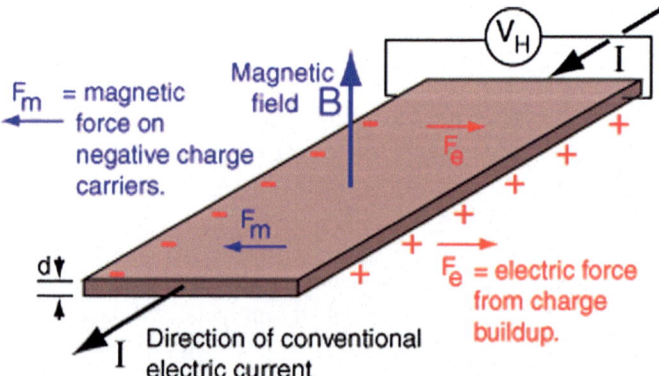

Fig. 1 The classical Hall effect. Adapted with permission from the HyperPhysics site, Department of Physics and Astronomy—Georgia State University; http://hyperphysics.phy-astr.gsu.edu

particular case of the previous one and occurs in few two-dimensional materials in which the electrons exhibit collective behavior with unexpected features.

In a 2D electron system in presence of a large perpendicular magnetic field and low temperatures, one can observe the integer quantum Hall effect, in which the Hall resistivity ρ_{xy} undergoes quantum Hall transitions whose values are quantized in integer units of h/e^2. This effect was discovered by von Klitzing [1] a century after the discovery of the classical Hall effect.

Exploration

In the Explore stage the students have the opportunity to get directly involved with phenomena and materials through observations and analysis of experimental results obtained in different hetero-structures. In this phase, the teacher acts as a facilitator, providing materials and guiding the students' focus. The Explore stage is the beginning of student involvement in inquiry. The learners search for information, raise questions, develop hypotheses to be tested, and collect data.

The experiments [2] show that ρ_{xy} varies by steps governed by the law

$$\rho_{xy} = h/se^2, \tag{4}$$

where s is an integer greater than zero that depends on B_z. The coefficient h/e^2 is a universal constant independent of the sample being employed.

Figure 2, obtained in a low-mobility two-dimensional electron gas in GaAs/Al$_x$Ga$_{1-x}$As with a carrier concentration $n = 1.93 \times 10^{11}/cm^3$ at a lattice temperature $T = 66$ mK, shows that the Hall resistivity ρ_{xy} (right hand scale in Fig. 2) is characterised by step behaviour with integer multiples of h/e^2. The magneto-resistance ρ_{xx} (whose scale is on the left) is appreciably different from zero when the Hall resistivity jumps from one plateau to another [2].

Explanation

The Explain stage is based on data and evidences processing techniques for the individual groups or the entire class (depending on the nature of investigation) from the information collected during the exploration. At this stage students formulate models (descriptive or explicative), discuss their data with peers and the teacher, and begin to argue with themselves what they have learned.

Stimulated by the teacher, the students realize that the IQHE is due to the fact that the closed circular orbits travelled by the electrons in a gas of independent (free) electrons can only occupy quantized energy levels [10], called *Landau levels*, that are solutions of the Schrödinger equation

$$E_n = (n + 1/2)\hbar\omega_c, \quad (n = 0, 1, 2, 3, \ldots) \tag{5}$$

The mechanism giving rise to the quantized Hall resistivity depends on how many of these energy levels are populated with electrons. The number of allowed orbitals on each orbit is constant for a given value of the magnetic field. The electrons cannot reside in the energy gaps in between or in the quasi-continuous k_z states (unaffected by B_z) that do not exist in a 2D crystal. Figure 3 shows how the transition from the quasi-continuous Fermi levels to the highly degenerate Landau levels occurs. The degeneration $\mathscr{D}$ increases with B_z according to the law $\mathscr{D} = \rho B_z$ with $\rho = eL_xL_y/h$ [10]. The population of the occupied Landau levels, consequently, depends on B_z (to be considered as a variable) and on the size of the conductor (a constant). The electrons occupy these levels in increasing order up to last one that may or may not be completely filled. This behaviour is characteristic of an *ideal* crystal structure.

In a *real* lattice, in which defects and impurities are unavoidable, things may be different. Suppose that only the lowest Landau level is completely filled and B_z is decreasing; the degeneration then also decreases, and an increasing number of electrons is compelled to migrate to the contiguous level that may be too high. From the energetic point of view, it will be more convenient for them to be trapped at localized impurity sites. Such electrons are not mobile and, consequently, they do not contribute to the conductivity. A further reduction of B_z reduces the Landau energy separation and it becomes convenient for the trapped electrons to occupy an available lower empty level. It turns out that the Landau levels in a real lattice are completely filled up or completely empty over a wide range of B_z. The plateau in each step of Fig. 2 is therefore explained in the context that there are always filled

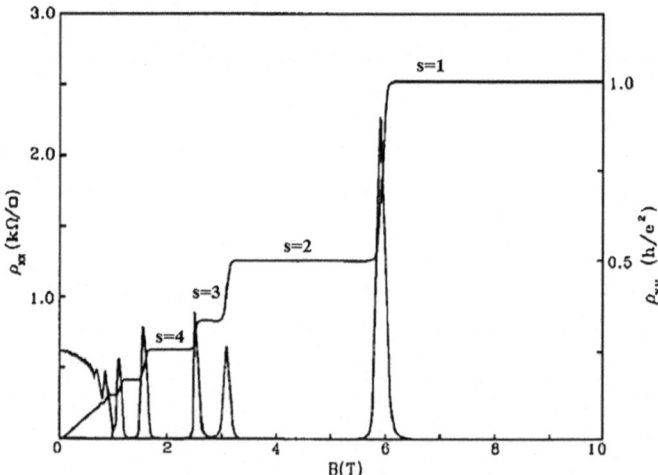

Fig. 2 ρ_{xx} and ρ_{xy} of a relatively low-mobility 2D electron gas in GaAs/Al$_x$Ga$_{1-x}$As; $T = 66$ mK, $n = 1.93 \times 10^{11}$/cm^3. Adapted with permission from Ref. [2]—Copyrighted by the American Physical Society

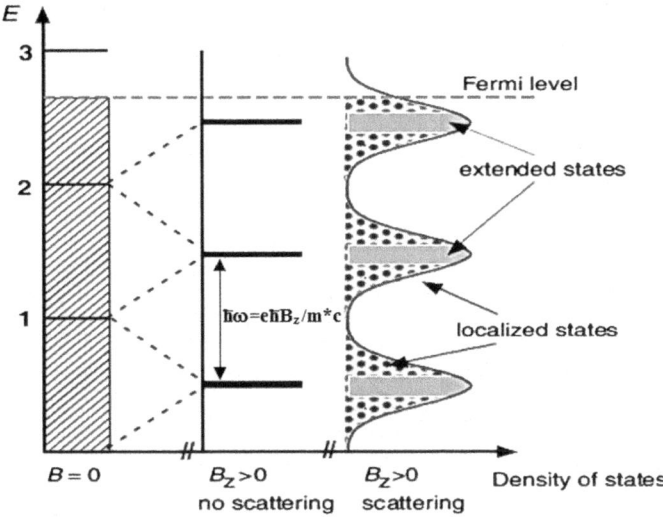

Fig. 3 Left: quasi-continuous states of a Fermi electron gas at T = 0 in absence of perturbations; center: rearrangement of the electron distribution on the left in presence of a magnetic field and absence of impurity/phonon scattering (Landau levels); right: rearrangement of the electron states in presence of impurities that cause scattering and electron trapping (localized states). Adapted with permission from [11]

Landau levels with the exception of those values of B_z that fall in narrow ranges in which a transition from the last occupied Landau level to the next empty one occurs. The magneto-resistance ρ_{xx} is consequently zero, since when some Landau levels are completely filled and others completely empty, there is no possibility of electron scattering. In the narrow transition region in which s assumes a contiguous value there are incompletely filled levels, scattering is possible, and consequently ρ_{xx} is not zero. It is very important that our students observe that in an ideal structure, without defects, IQHE is not possible.

Elaboration/Extension to the FQHE

In the Extend stage students expand the concepts they have learned, make connections to other related concepts, and apply their understandings to the world around them in new ways, building possible generalizations.

The fractional quantum Hall effect, a particular case of the quantum Hall effect, occurs in a few 2D materials such as graphene in which the electrons exhibit collective behaviour with unexpected features. This effect was discovered by Tsui and Stormer in 1981 in 2D hetero-junctions in which the lowest Landau level is partially occupied because of the low electron density [2]. The FQHE is detectable in these materials at very low temperatures (<1 K) and extremely high magnetic fields (>150 T). The main difference between the IQHE and the FQHE consists in the basic property of the electrons that behave as free (non-interacting) particles in IQHE and correlated interacting particles in the second case.

The equal spacing of the Landau levels is here replaced by a collective behaviour of *composite* particles whose charge is smaller than e and the level separation is proportional to $B_z^{1/2}$. This effect was not expected since the charge of other composite particles, such as the Cooper pairs in superconductors, is a multiple of e. It has been explained in the context that bits of magnetic field are attached to each electron creating a new object whose properties are different from those a free or a bound electron [16]. These particles experience an effective magnetic field B_z^*, different from the applied field B_z. Their motion seems do not depend strictly on the magnetic field and they may behave as bosons as well as fermions depending on the value of the magnetic field [12]. The experimental results found by Novoselov et al. [13] in graphene are reported in Fig. 4. The transversal conductance σ_{xy} (along the y-direction) is quantized in steps with half integer multiples of $4e^2/h$ (right hand scale in Fig. 4) and can assume positive as well as negative values ascribed to the electrons whose energy is greater than zero with respect to the Fermi level or to the hole conductivity whose energy is negative, respectively. The longitudinal resistivity ρ_{xx}, whose scale is on the left, is close to zero in each plateau interval.

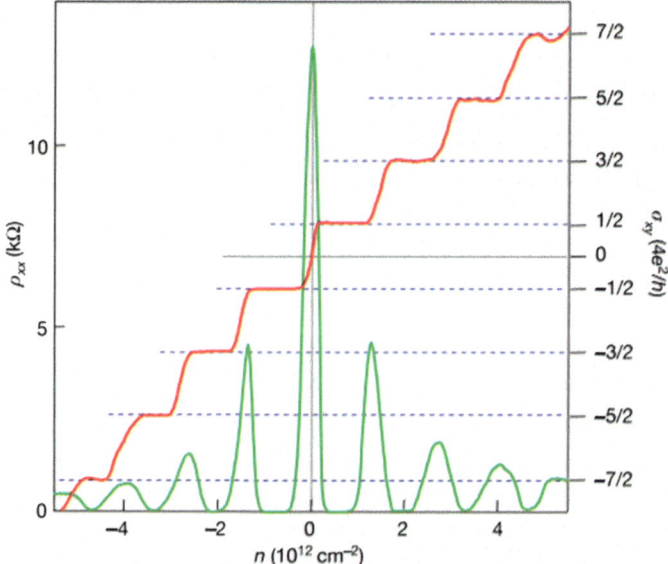

Fig. 4 Hall conductivity σ_{xy} and longitudinal resistivity ρ_{xx} of graphene as a function of their concentration at $B = 14$ T and $T = 4$ K. Adapted with permission from Ref. [13]—Copyrighted by Nature Publishing Group

The major limit to commercial applications of the quantum Hall effects is inherent to the need of extremely low temperatures (less than 1 K) and high magnetic fields. However, Novoselov et al. [14] have pioneered a research having the aim of avoiding the first difficulty. They were successful in showing that in single layer graphene the quantum Hall effect can be observed at room temperature. This is possible for various reasons. The most relevant are the unusual nature of the charge carriers in this material, they behave as massless relativistic particles whose motion is little affected by scattering, and their high concentration [6].

Evaluation

The Evaluation stage is an on-going diagnostic process for both students and teachers. It involves students' capacity to make judgments, analyses, and evaluations of their work, also in comparison with the work of their colleagues. It also allows teachers to determine how much learning and understanding have taken place.

Both quantum Hall effects are rare observable examples of microscopic effects on a macroscopic scale and allow us to establish very precise values of microscopic quantities such as the electron charge and the Planck constant that can be measured without being disturbed by the influence of the materials being used or by the combination with other unknown quantities. The quantum Hall effect results from an interesting interplay between disorder, and interactions, but the basic physics is that of an electron in a magnetic field. Moreover, it evidences, in specific cases, the existence of interaction processes that appear as a fractional charge and helps us to understand some physics governing interacting particles.

At the end of the workshop the learners should have consolidated their knowledge of the Hall effect and achieved a more meaningful understanding of relevant concepts regarding the quantum behavior of correlated particles [3]. In particular they should be able to look at the phenomenon from a general viewpoint by making analogies with other atypical behaviors such as Cooper pairs in superconductor materials and composite electrons in graphene, and appreciate analogies and differences between the concepts of the effective mass of electron in semiconductors and the effective charge of Dirac fermions in graphene.

Conclusion

The quantum Hall effect in graphene, and Cooper pairs in superconductors represent two of the very few quantum effects observable at macroscopic level. In both cases, the correlated behavior of the particles gives rise to effects not observable when we are dealing with processes typical of independent or weakly interacting particles, such as free electrons in metals or nearly free electrons in semiconductors. The 5E-based workshop presented here stimulates a discussion about the different features of the

Hall effects (classical, integer and fractional) detectable under various and sometimes extreme experimental conditions, by highlighting their dissimilar physical grounds. At the basis of the learning experience is the idea of blending and interconnecting separate pieces of knowledge already acquired by undergraduates in different courses and to help them to visualize and link the concepts lying beyond separate chunks of information or equations.

Our educational environment, focused on an active student engagement and the application of the concepts previously introduced within a traditional lecture-based instruction context, could provide the students with a valuable experience useful for the development of long-lasting skills [15].

References

1. von Klitzing, K.: The quantized Hall effect. Rev. Mod. Phys. **58**, 519-531 (1986). https://doi.org/10.1103/RevModPhys.58.519
2. Stormer, H.L.: Nobel lecture: the fractional quantum Hall effect. Rev. Mod. Phys. **71**, 875–889 (1999). https://doi.org/10.1103/RevModPhys.71.875; Tsui, D.C.: Nobel lecture: interplay of disorder and interaction in two-dimensional electron gas in intense magnetic fields. Rev. Mod. Phys. **71**, 891–895 (1999). https://doi.org/10.1103/RevModPhys.71.891
3. Johnson, B.L., Kirczenow, G.: Composite fermions in the quantum Hall effect. Rep. Prog. Phys. **60**(9), 889–939 (1997). https://doi.org/10.1088/0034-4885/60/9/002
4. Novoselov, K.S.: Nobel lecture: graphene: materials in the flatland. Rev. Mod. Phys. **83**, 837–849 (2011). https://doi.org/10.1103/RevModPhys.83.837
5. Persano Adorno, D., Bellomonte, L., Pizzolato, N.: Electronic properties of graphene: a learning path for undergraduate students, chap. 18. In: Greczyło, T., Dębowska, E. (eds.) Key Competences in Physics Teaching and Learning. Springer Proceedings in Physics, vol. 190, pp. 215–227 (2017). https://doi.org/10.1007/978-3-319-44887-9
6. Persano Adorno, D., Bellomonte, L., Pizzolato, N.: The amazing graphene: an educational bridge connecting different physics concepts. Eur. J. Phys. **39**(1), 013001 (2018). https://doi.org/10.1088/1361-6404/aa91a3
7. Persano Adorno, D., Pizzolato, N., Fazio, C.: Elucidating the electron transport in semiconductors via Monte Carlo simulations: an inquiry-driven learning path for engineering undergraduates. Eur. J. Phys. **36**(5), 055017 (2015). https://doi.org/10.1088/0143-0807/36/5/055017
8. Kroll, L.R., Laboskey, V.K.: Practicing what we preach: constructivism in a teacher education program. Action in Teacher Education **18**(2), 63–72 (1996). https://doi.org/10.1080/01626620.1996.10462834
9. Bybee, R.: Achieving Scientific Literacy: From Purposes to Practices. Heinemann, Portsmouth NH (1997)
10. Kittel, C.: Introduction to Solid State Physics, 8th edn. Wiley, New York (2005)
11. Longo, M.: Effetto Hall su graphene. Tesi di laurea in Fisica, Università degli Studi della Calabria (2011)
12. Laughlin, R.B.: Anomalous quantum Hall effect: an incompressible quantum fluid with fractionally charged excitations. Phys. Rev. Lett. **50**(18), 1395–1398 (1983). https://doi.org/10.1103/PhysRevLett.50.1395
13. Novoselov, K.S., Geim, A.K., Morozov, S.V., Jiang, D., Katsnelson, M.I., Grigorieva, I.V., Dubonos, S.V., Firsov, A.A.: Two-dimensional gas of massless Dirac fermions in graphene. Nature **438**, 197–200 (2005). https://doi.org/10.1038/nature04233
14. Novoselov, K.S., Jiang, Z., Zhang, Y., Morozov, S.V., Stormer, H.L., Zeitler, U., Maan, J.C., Boebinger, G.S., Kim, P., Geim, A.K.: Room-temperature quantum Hall effect in graphene. Science **315**(5817), 1379 (2007). https://doi.org/10.1126/science.1137201

15. Persano Adorno, D., Pizzolato, N., Fazio, C.: Long term stability of learning outcomes in undergraduates after an open-inquiry instruction on thermal science. Phys. Rev. Phys. Educ. Res. **14**, 010108 (11 pp.) (2018). https://doi.org/10.1103/physrevphyseducres.14.010108

The Broken Mirror: The Foundations of Thermodynamics and the Failure of Mathematics to Reflect the Physics

David Sands(iD)

Abstract The art of the physicist is to express a physical situation mathematically and conversely, interpret the results of a mathematical analysis in terms of the physics. Physics should lead the mathematics. Yet, for many students, a formula is just a tool to solve a problem. Not surprisingly, solving a problem becomes an exercise in finding the right tool and physics takes a back seat. Even among professional physicists there can be a tendency to put mathematics first and physics second. However, a mathematical result that appears to agree with physics is not necessarily correct, as there might be assumptions along the way that are unphysical. In this paper I trace the origins of the concept of entropy laid down by Clausius in the middle of the nineteenth century. I will show how Clausius failed to link the mathematical structure he was developing to physical processes and consider the implications for our understanding of the Second Law. Finally, I address the question: if the theory of thermodynamics is as flawed as I believe it is, what should we be teaching our students? How can we encourage students to examine critically the connection between maths and physics and at the same time teach a theory in which this connection appears to be missing?

Keywords Entropy · State function · Classical ideal gas · Reversible · Irreversible

Introduction

Clausius and Carnot are two of the most important figures in the history of thermodynamics: Carnot, for his idealization of a heat engine that paved the way for people like Clausius and Thompson (Kelvin), and Clausius for his reworking of Carnot's theory to take into account changing theories of heat. The Carnot cycle is thus the foundation upon which thermodynamics has been built. In 1850, Clausius restructured Carnot's ideas to accord with the prevailing idea of heat-as-motion and subsequently developed the modern notion of reversibility from the idea of a

D. Sands (✉)
Department of Physics and Mathematics, University of Hull, Hull, UK
e-mail: d.sands@hull.ac.uk

reversible cycle. Considerations of reversibility and irreversibility have been prominent in thermodynamics ever since, such that modern thermodynamics is divided
into two separate branches dealing with these very issues: classical, or reversible,
thermodynamics dealing with equilibrium states, and irreversible thermodynamics
dealing with transient or dynamic phenomena. This situation is due almost entirely to
Clausius. In fact, had it not been for Clausius, Carnot's work might well have slipped
into obscurity. Two years after Carnot's death in 1832, Clapeyron published what
Porter [1] describes as "a more elegant" account of Carnot's work without adding to
it, but by 1845, when Kelvin visited Paris, no "bookseller had even heard of Carnot".
When Clausius published his reconstruction of Carnot's theory in 1850 he not only
dispensed with the now defunct idea that heat was a material substance but also incorporated Joule's idea on the equivalence of work and heat and, according to Porter, at
once removed any doubts that Kelvin had over Carnot's work. Shortly after, Kelvin
published his own version of the second principle of thermodynamics, now known
as The Second Law, acknowledging that, "The whole theory of the motive power of
heat is founded … on two propositions due respectively to Joule and to Carnot and
Clausius" [1].

Somewhat perversely, though, Clausius' ideas also made the idea of a heat engine
seem something of an abstraction, for although both Carnot and Clausius used the
word "motive" to describe the power of heat to do work, Clausius explicitly required
the engine to run quasi-statically, which means that the engine cannot develop power.
In effect the ideal engine cannot function as an engine. Perhaps it is this rather than any
intrinsic difficulty with the concept of the Carnot cycle that makes its relationship
with heat engines difficult for students to comprehend. This raises the intriguing
possibility that Clausius misunderstood the Carnot cycle and indeed, it is possible
to see in Clausius' writings a direct connection with Carnot's ideas on caloric that
survived his reconstruction of Carnot's theory. I suggest here that in fact, Clausius'
ideas on heat did not correspond to the modern understanding of the term and that
not only were the consequences of this not fully realized at the time, but modern
thermodynamics is flawed in consequence. I start by tracing Clausius' development
of the concept of entropy and then discuss, using some simple examples, the kind of
difficulties that can arise in thermodynamics. I finish by looking at the implications
for the teaching of thermodynamics.

The Development of the Concept of Entropy

It is possible to see in mid-19th century writings a transitional view of heat, but it
is neither explicit nor obvious. The modern reader is able to understand much with
no more than the odd moment of confusion, but a slight shift in perspective can
remove the confusion altogether. Take, for example, Tyndall's Heat Considered As
A Mode Of Motion, first published in 1863 [2], around the same time as Clausius
was writing his Sixth Memoir [3] in which he extended the Second Law to non-
cyclic processes. Tyndall describes lifting a lead weight a height of sixteen feet and

letting it drop. At this time the notion of energy was still under development and Tyndall discussed "possible" or "potential" energy as opposed to "moving force" or "actual" energy. Tyndall discusses how much heat would be required to raise the weight against gravity and in so far as thermodynamics was at that stage concerned with the conversion of heat into work through the medium of the heat engine, there would appear to be nothing odd about this. However, Tyndall doesn't mention heat engines, but writes that to raise the weight sixteen feet would "consume as much heat as would raise the temperature of a cubic foot of air about 1 °F. Conversely, this amount of heat would be generated by the falling of the weight from a height of sixteen feet". The implication is that the heat is represented by the motion itself.

Indeed, this is a recurring theme. In the preface, Tyndall describes the various forms of what he calls "thermometric heat": heat that can be sensed using a thermometer, and includes among the various forms, "the conception of heat as molecular motion". When writing about heating bodies through friction, Tyndall writes that "... the heat produced is the measure of the force expended in overcoming the friction. The heat is simply the primitive force in another form ...". From the modern perspective, the identification of heat with force is incorrect, but the concept of energy hadn't been fully developed by this time. For example, the motive power to which Carnot referred was not power as we understand the term, which is rate of conversion of energy, but work done or perhaps even the potential to do work. There is a looseness in the terminology which refinements in the concept of energy eliminated, but at the time Clausius was writing it appears to have been common to equate heat with motion and what we now call kinetic energy and attribute to it a force.

Tyndall also described the notion of heat being consumed during internal work, by which is meant the separation of atoms against attractive interparticle forces. This topic was of great interest to Clausius and formed the motivation for the development of the concept of entropy. Tyndall's view is not very far removed from the current view that the effect of heating is to bring about increased vibration in solids or increased velocity of the molecules in gases, but Clausius is susceptible to a very different interpretation. He writes in his Sixth Memoir that it is "... 'self-evident', that heat [kinetic energy] must be consumed in the production of internal work and, vice versa, generated if internal work is expended ...". This can only be understood if heat is equated with motion and kinetic energy.

This view of heat and its connection to internal work was central to Clausius' conception of entropy. In treating internal work, Clausius borrowed from his earlier work on cyclic processes: "... as there is no essential difference between interior and exterior work, we may assume with certainty that a theorem which is so generally applicable to exterior work cannot be restricted to this alone". Exterior work is that produced by a heat engine and interior work is that done against inter-particle forces. The production of work from heat was not fully understood. It was seen as some sort of transformation that took place within the gas and Clausius' statement about a theorem developed for exterior work applying to interior reflects this. In consequence, he took a view of thermodynamics developed for cyclic processes and applied it to non-cyclic processes.

There are, in fact, two theorems in the mechanical theory of heat, which Clausius explains in the Ninth Memoir [4] as the equivalence of heat and work, or Joule's principle, and the equivalence of transformations developed by Clausius. It was to this theorem that Clausius referred, but the modern reader will be unfamiliar with the idea. It is an obscure concept which has disappeared altogether from the modern view of thermodynamics. The two transformations which Clausius regarded as equivalent in the were explained in the Sixth Memoir as the conversion of heat into work, and vice versa, in a cyclic process and the conversion of "heat at one temperature to heat at another temperature". This last idea is almost a direct echo of Carnot and illustrates the transitional nature of Clausius' thinking.

Mathematically, the theorem of the equivalence of transformations can be expressed as

$$\int \frac{dQ}{T} \geq 0 \qquad (1)$$

The positive sign arises because Clausius regarded Q as positive when heat is produced from work and therefore "given up by the body". Clausius had written in the Sixth Memoir that although "this theorem (Eq. 1) admits of strict mathematical proof … it … nevertheless retains an abstract form in which it is with difficulty embraced by the mind, and we feel compelled to seek for the precise physical cause …". In fact, from the modern perspective the interpretation of Eq. (1) is entirely straightforward, but Clausius' view of heat was different from ours. Despite this lack of understanding, Clausius asserted in the very next sentence that a theorem applying to external work must apply to internal work and actively sought an inequality of the form of Eq. (1) that characterized non-cyclic processes. It is my contention that in so doing, he violated the law of conservation of energy, but the inequality he sought has been carried forward today in the form of the now famous inequality of irreversible thermodynamics,

$$T\,dS \geq dQ \qquad (2)$$

The contradiction of the First Law implied by this has never been recognized and the lack of a physical mechanism underlying it is the principal reason for the title of this paper.

The difficulty that Clausius faced in trying to establish an equivalent inequality for non-cyclic processes was the complete lack of knowledge of internal forces. Clausius therefore turned his attention to work: "This difficulty disappears if we take into calculation, not the forces themselves, but the mechanical work which, in any change of arrangement, is required to overcome them". Borrowing directly from a law he had formulated in relation to external work, Clausius developed a similar law for internal work. For external work, he had written: "In all cases in which the heat contained in a body does mechanical work by overcoming resistances, the magnitude of the resistances which it is capable of overcoming is proportional to the absolute temperature". For internal work, the equivalent law was: "The mechanical work which can be done by heat during any change of arrangement of a body is proportional to the absolute temperature at which this change occurs."

The modern reader would probably interpret the first statement as a rather cumbersome way of saying that pressure depends on absolute temperature, but Clausius refers elsewhere to "the force of the heat" and appears to regard it as something different from pressure: "... the externally sensible pressure of a gas forms an approximate measure of the separative force of the heat contained in a gas; and hence ... this pressure must be nearly proportional to the absolute temperature". The theorem on internal work concerning the "mechanical work which can be done by heat" clearly implies that Clausius believed heat to possess in some manner a force, the magnitude of which depended on the absolute temperature. Clausius was interested in the potentiality of that force: "The law does not speak of the work which the heat *does*, but of the work which it *can do* ...". The emphasis is Clausius', but he applied the same reasoning to both internal and external work: "... similarly, in the first form of the law, it is not of the resistances which the heat overcomes, but those of which it *can overcome* that mention is made".

It is easy to overlook the significance of this. It comes from equating heat with kinetic energy and the idea that heat possesses a force which has to be measured: "In order ... to determine the force of the heat, we must evidently not consider the resistance which actually is overcome, but that which can be overcome". By putting the emphasis on the work that could be done rather than that which is done, Clausius has in effect dispensed with energy conservation as defined in the First Law, which is concerned solely with changes that have actually occurred. The consequences of this for entropy will become clear as we go.

In order to illustrate how "the force of the heat" could be measured in terms of work that could be done, Clausius considered the case of a gas expanding into vacuum or into a space at a lower pressure: the expanding gas overcomes a resistance less than it would be possible for it to overcome. This is an irreversible process in as much as work must be done on the gas to restore the initial state. Were the initial state to be restored, the inequality in Eq. (1) would apply. The meaning of this inequality now becomes apparent. In an irreversible cycle the initial state can only be achieved by a net outflow of heat such that $\delta Q/T$ taken around the cycle is non-zero. As Clausius defined it, the sign is positive but in modern usage it is negative. This is entirely general. In the specific case of the free expansion, heat must be extracted following the work done to compress the gas without any corresponding influx of heat during the expansion. There is no conflict with the First Law. In fact, the inequality is required by the First Law because to return to the initial state any work done on or by the system must be counteracted by a corresponding heat flow.

In a single non-cyclic process, on the other hand, work and heat cannot be separated in this way. The quantity $\delta Q/T$ can be positive or negative as heat flows into or out of a system, but the First Law requires the sum of the changes in internal energy and work done to match the heat flow. Therefore, it is not clear what is changing in the system that has the units of energy and is *always* greater than, or equal to, the change in heat (Eq. 2). Yet, Clausius believed that such a quantity exists and that it is a property of a substance in a given thermodynamic state.

In order to develop this idea mathematically, Clausius introduced the concept of disgregation, Z. A reversible process was defined by Clausius thus: "When a change of arrangement takes place so that force and counterforce are equal, the change can likewise take place in the reverse direction under the influence of the same forces." For reversible processes, TdZ is equal to the sum of internal and external work, but for an irreversible external process, TdZ is given by the sum of internal work and PdV, the latter representing the external measure of "the force of the heat". Thus, TdZ is simply the magnitude of the work *that could be done* (my emphasis) if the same transformation were achieved reversibly. Clausius therefore arrived at the following expression for an infinitesimal change:

$$dQ + dH + TdZ \geq 0 \tag{3}$$

Here dQ is the external heat exchanged with the environment and dH is the change in "the quantity of heat present in the body", with H itself being considered by Clausius to be proportional to the absolute temperature. In effect, H is the kinetic energy of the constituent particles. The sum, $dQ + dH$, is the "new quantity of heat which ... must be produced from work or ... converted into work" and is positive if heat is produced and negative if heat is converted into work. If the equality holds, the process is reversible and the quantity of heat matches the maximum amount of work that can be done. In the language of Clausius, the transformations are compensated. If the inequality holds, the change in heat is algebraically less than TdZ and there is an uncompensated transformation. This is the result that Clausius sought and which, in his view, put non-cyclic processes on the same footing as cyclic processes. The difficulty with it, and it is worth emphasizing again, is that it is not based on changes that actually occur, which are governed by the First Law, but on changes which could occur if, to use Clausius' phrasing, the force of the heat were to do the maximum work it is capable of doing.

Equation (3) is not recognizable as the modern expression of entropy. It was not until the Ninth Memoir of 1865 that Clausius coined the term "entropy". In this Memoir, Clausius restated the two fundamental theorems of the mechanical theory of heat: the equivalence of heat and work and the equivalence of transformations, but for the latter he adopted the opposite sign convention from Eq. (1), with the recognition that a quantity of heat absorbed by a body is positive and a quantity of heat given up is negative. Therefore, the equivalence of transformations became, for larger changes,

$$\int \frac{dQ}{T} \leq 0 \tag{4}$$

The modern form of entropy was introduced as:

$$dS = \frac{dQ}{T} \tag{5}$$

For reversible processes,

$$S = S_0 + \int \frac{dQ}{T} \tag{6}$$

For irreversible processes,

$$\int \frac{dQ}{T} = \int \frac{dH}{T} + \int dZ \tag{7}$$

The term on the right was called by Clausius, the entropy of a body.

In summary, there are several inter-related ideas leading up to Eq. (7) that find no equivalent in modern thermodynamics but are crucial to the concept of entropy as it was first developed and still echo through the structure of thermodynamics today. First, and most obviously, there are the two quantities that make up the entropy, H and Z, which represent together the idea that a body contains a definite amount of "actual heat" (H) at a given temperature and that this heat possesses a force that can be measured by TdZ, or the work done in a reversible process. Crucially, the transformation from heat to work and vice versa was believed by Clausius, and perhaps others among his contemporaries, to occur within the body and in consequence was regarded as a property of a body. This idea of transformation occurring within a body is central to the concept of entropy: "We might call S the *transformational content* of the body … [but] … I propose to call the magnitude S the *entropy* of the body, from the Greek word $\tau\rho o\pi\eta$, *transformation*".

Entropy in Modern Thermodynamics

Modern thermodynamics does not recognize any of the ideas on which Clausius definition of entropy rests, but it is still essentially his ideas that apply: that is, entropy is regarded as a property of a body (a state function) which increases during an irreversible process. The idea that entropy is a state function has a particular mathematical interpretation which will be addressed in due course, but there is another vital aspect of entropy that has emerged since Clausius: the connection with statistical mechanics.

The first demonstration of this connection came late in the 19th century with Boltzmann's H-theorem [5], his mathematical proof that Maxwell's velocity distribution is the equilibrium distribution of velocities in an ideal gas. In a remarkable mathematical *tour de force*, Boltzmann arrived at the Maxwellian probability distribution p by minimizing the function

$$H = \sum p \ln p \tag{8}$$

This turns out to have the same mathematical form as the thermodynamic entropy apart from a multiplying constant, to which Boltzmann's name was later attached. The Maxwellian distribution has the same form as Gibbs' canonical distribution in statistical mechanics [6] so the H-function of Eq. (8) also has the same form and is identical to the thermodynamic entropy aside from the multiplying constant. It soon became apparent, however, that thermodynamic entropy and statistical entropy were not identical under all circumstances, as shown by the so-called Gibbs paradox.

The Gibbs Paradox and Irreversible Expansion

Entropy is considered to be an extensive property, which can be expressed mathematically by representing it as a homogeneous function of degree 1. This has the property that if all the variables on which the function depends are increased by the same factor, the function itself is increased in direct proportion: that is, if S is a function of U, V and N, then

$$S(\alpha U, \alpha V, \alpha N) = \alpha S(U, V, N) \tag{9}$$

If, then, we consider two equal volumes, V, of equal amounts (N) of an ideal gas at the same temperature (equal U) in adjacent compartments of the same container separated only by a removable partition, removal of the partition changes the system from two chambers to one at twice the volume with twice the number of particles and twice the internal energy. This is equivalent to setting $\alpha = 2$ in Eq. (9): the entropy of the whole will be twice the entropy of each individual chamber prior to the removal of the partition. As there are two initial chambers, each with half the entropy of the final, there is no entropy change on removal of the partition. However, the statistical entropy depends on $Nk\ln V$ and therefore increases by a factor of $2Nk\ln 2$.

The most common resolution of this difficulty is to assume that the particles are indistinguishable and that the states of the system are overcounted as a result. Dividing the partition function by $N!$ introduces a term $-Nk\ln N$ into the entropy and effectively normalizes the volume to the number of particles. Therefore, the entropy remains unchanged if the gases on either side of the partition are identical, but increases if they are different.

This changes the definition of statistical entropy from a distribution over particles to a distribution over arrangements, but it leads to another difficulty. The increase in entropy is often attributed to mixing, but in fact it is due to irreversibility: essentially, each gas is effectively expanding into an empty volume. Imagine instead of two chambers each with N particles, one chamber with $2N$ particles adjacent to an empty chamber. Removal of the partition will cause a free expansion and increase the entropy, regardless of whether the gas comprises identical particles or N_A of gas A and N_B of gas B mixed together. However, this introduces a conundrum. It is clear that if the initial gas is a mixture, the entropy of $N_A + N_B$ in a volume V is just the sum of the entropies of N_A and N_B contained separately in V, but for like particles the entropy of $2N$ particles in V is less than twice the entropy of N particles in V. In other words, the entropy of distinguishable particles is additive whilst that of indistinguishable particles is not.

As described at length above, the idea that entropy changes in an irreversible expansion is a direct consequence of Clausius' idea that what is important is the work that can be done, or the reversible work, PdV, integrated over the volume change. It is here that the conflict with the First Law is laid bare: PdV is a work term, but no work is done during the process. In consequence, TdS, which has the units of energy, increases according to Eq. (2) when there are no energy changes in the system. The implications of this for energy conservation, and by implication the standing of the law of entropy increase relative to the First Law, has never been examined. In mathematical terms, a quantity, say $d\varphi$, can always be added to dQ to bring about the equality, but the physical origins of $d\varphi$ are open to question as dQ and its relationship to dU and dW is defined by the First Law.

One possible reason why this aspect of entropy has never been questioned might be that entropy is seen to be a state function and in consequence must increase during an irreversible transition. Both dQ and dW are path-dependent variables, but an integrating factor of $1/T$ converts them to path-independent variables. The quantity $dS = dQ/T$ is therefore an exact differential and is recognized as the entropy. However, it is implicit in the mathematical process of integration that both dQ and dW are infinitesimal changes, which raises the question as to whether there are any physical processes that correspond to this mathematical abstraction. In practice it is likely that a finite change of constraint will be required to bring about a change of state, a situation recognized by Clausius but disregarded.

Concentrating on work rather than heat flow, suppose a very small weight, δW, is added or removed from a piston in order to initiate an adiabatic process and suppose further that the weight can be made arbitrarily small. Does this ever approach PdV or does it always require a step-wise change? As a process it is clearly irreversible: the weight acts during compression but is removed for expansion. However, if there is a sense that in the limit of ever smaller changes the system returns to the same state if the external constraint is restored, we can conceive of a physical process that matches the mathematics and regard it as intrinsically reversible. If not, the mathematical process of integration does not correspond to any physical process and we would be left with the possibility that the notion of entropy as a state function is no more than a mathematical construct that has no correspondence with any physical process.

In fact, there is no intrinsic reason to suppose that the initial state is restored or that work is reversible. The thermodynamic state of an ideal gas is defined by P, V and T, but for an adiabatic process such as that described above both T and V can vary during the process and only P is defined by the external conditions. In short, restoration of the initial pressure requires that only the ratio $T : V$ must be restored. According to Eq. (4), if the process is irreversible it would be necessary to extract heat from the system in order to return to the initial state. It shouldn't matter whether we consider expansion followed by compression or compression followed by expansion, an irreversible process would require the temperature of the gas after this two-stage process to exceed the initial temperature in order to allow heat to be extracted. Expansion would therefore require a larger volume change than compression.

The first issue to be resolved is the nature of damping. For a finite change of pressure, no matter how small, the system would oscillate, and some mechanism of damping would be required. If the damping is external the energy change in the gas could be the same in both directions, but in fact an internal mechanism of damping exists [7], which means that we can idealize to a frictionless piston and still consider a definite change in volume. Suppose the external pressure during expansion is P_e and the volume change is δV_e. The work done by the gas is $P_e \delta V_e$, which is just the external work required to lift the piston. In the reverse direction, denoting compression by the subscript c, the work done by the piston is $P_c \delta V_c$. The absence of any external damping means that the kinetic energy of the piston is dissipated within the gas and the external work corresponds to the energy changes within the gas. It follows that if the external pressure is changed and then restored, say by adding and removing a small weight in whatever order, then $\delta V_e > \delta V_c$. If, instead, $\delta V_e \leq \delta V_c$ the work done on compression would be obviously greater than that done by expansion by virtue of both a greater pressure and a greater volume change. Therefore, on restoration of the initial pressure the internal energy in the gas would be higher than the initial temperature, but with a volume either smaller than or the same as the initial volume, which is obviously impossible. It follows, therefore, that the volume change on compression *must* be smaller than on expansion. Therefore, the final state at the end of a two-stage process corresponds to a higher temperature and larger volume, the ratio of which gives the initial pressure. Therefore, heat has to be extracted from the system to reduce the volume and restore the initial state and the process is intrinsically irreversible.

This argument is supported by the energy changes within the gas. Identifying two slightly different state by subscripts 1 and 2, a small change of external constraint gives rise to a difference in internal energy between the two states of:

$$\Delta U = \tfrac{3}{2}(P_2 V_2 - P_1 V_1) \tag{10}$$

If one state is written in terms of the other, say $P_1 = P_2 - \delta P$ and $V_1 = V_2 + \delta V$, then:

$$\Delta U = \tfrac{3}{2}(\delta P V_2 - P_2 \delta V + \delta P \delta V) \tag{11}$$

The first quantity on the right, $\delta P V_2$, can be found as follows. As this is an adiabatic process, the states (P_1, V_1) and (P_2, V_2) are linked by the relationship

$$P_2 V_2^\gamma = P_1 V_1^\gamma = (P_2 - \delta P)(V_2 + \delta V)^\gamma \tag{12}$$

The last term on the right can be re-written using the binomial expansion as:

$$(V_2 + \delta V)^\gamma = \left[V_2 \left(1 + \frac{\delta V}{V_2} \right) \right]^\gamma \approx V_2^\gamma \left(1 + \frac{\gamma \delta V}{V_2} \right) \tag{13}$$

Substituting back into Eq. (12) and expanding the brackets yields;

$$\delta P V_2^\gamma = \gamma P_2 V_2^{\gamma-1} \delta V - \gamma \delta P V_2^{\gamma-1} \delta V \tag{14}$$

Dividing through by $V_2^{\gamma-1}$ yields,

$$\delta P V_2 = \gamma P_2 \delta V - \gamma \delta P \delta V \tag{15}$$

Equation (11) then becomes,

$$\Delta U = \tfrac{3}{2} \left[P_2 \delta V (\gamma - 1) + \delta P \delta V (1 - \gamma) \right] \tag{16}$$

Therefore,

$$\Delta U = P_2 \delta V - \delta P \delta V = P_1 \delta V \tag{17}$$

Equation (17) is asymmetric. In Eq. (11), P_1 is expressed in terms of P_2 and likewise with the volume, but had it been the other way around the work done would have come to:

$$\Delta U = P_1 \delta V + \delta P \delta V = P_2 \delta V \tag{18}$$

The two values of the change in internal energy in Eqs. (17) and (18) correspond to the external work done in expansion and compression respectively and therefore support the preceding qualitative argument that a finite change is intrinsically irreversible. The only approximation is that the change is small enough for a first order binomial expansion to apply, so in principle the smaller the change the more accurate the result. Even in the limit as $\delta V \rightarrow dV$ the asymmetry between the two expressions will still exist. Adiabatic work would therefore appear to be intrinsically irreversible such that even for an infinitesimally small change heat must be extracted to restore the original state. This implies, then, that there is no work process corresponding to reversible work, PdV, central to the mathematical development of the concept of entropy.

Discussion and Conclusion

It has been argued in this paper that Clausius' conception of entropy was flawed, being based around the concept of transformations rather than conservation of energy. This led him to look for what he called an "uncompensated transformation" in a non-cyclic process such that a similar inequality to that applying to cyclic processes could be derived. He compared the work that is actually done to the work that could be done in a so-called reversible process and thereby introduced an inequality into the First Law (Eq. 3) that has been passed down to the present time in the form of Eq. (2), the famous inequality for irreversible thermodynamics. By incorporating a term for work that is not actually performed during an irreversible process, Clausius has arrived at a quantity, TdS, that appears to be a property of a body, has the units of energy, but increases in a manner not consistent with the energy changes governed by the First Law. In consequence, the idea that the entropy of a body increases in irreversible adiabatic process does not seem to be consistent with energy conservation.

In addition, it has been argued here that adiabatic work processes are intrinsically irreversible in an ideal gas in as much as the work of compression over a given volume change will always exceed the work of expansion and any finite change of state would not appear to be reversible. There are two implications of this. First, as discussed in this paper, the notion of entropy as a state function appears to be a mathematical abstraction not matched in real, physical processes. Secondly, and not touched upon in this paper for lack of room, Clausius' conception of the Carnot cycle, which relies on the adiabatic stages being conducted reversibly, is also flawed.

It is the author's belief that Clausius fundamentally misunderstood Carnot, who was concerned with the reversibility of the cycle itself. Performed in one direction, the cycle converts heat into work and performed in the other it converts work into heat. This was also the view of Kelvin, but history has favored Clausius. The implication of the work summarized here is that Kelvin was correct. There is a powerful argument, therefore, for returning to the origins of the subject and basing thermodynamics education not on entropy as a driving force or a determinant of equilibrium, but on cyclic processes. This would require a significant revolution in thinking about the subject and in consequence is far too complicated a topic to discuss in any kind of detail within this paper, but there is an important and immediate question for physics educators around the globe: if the theory of thermodynamics found in text books and almost universally taught in universities is flawed in that the mathematical structure does not reflect the physical processes, how should the physics education community respond?

References

1. Porter, A.W.: Thermodynamics, Methuen's Monographs on Physical Subjects, Methuen, London (1931)
2. Tyndall, J.: Heat Considered as a Mode of Motion. Longman, Green, Longman, Roberts and Green, London (1863)
3. Clausius, R.: The Mechanical Theory of Heat with Its Applications to the Steam Engine and to the Physical Properties of Bodies. In: Hirst, T.A. (ed.), p. 215. Taylor & Francis, London (1898)
4. Clausius, R.: Op. Cit., p. 327
5. Boltzmann, L.: Further studies on the thermal equilibrium of gas molecules. In: Brush, S.G., Hall, N.S. (eds.) History of Modern Physical Sciences. The Kinetic Theory of Gases: an Anthology of Classic Papers with Historical Commentary, vol. 1, pp. 262–349. World Scientific (2003)
6. Kittel, C.: Elementary Statistical Physics. Wiley, New York (1958)
7. Sands, D., Dunning-Davies, J.: Thermal damping in the compound piston. J. Non-Equilib. Thermodyn. **35**, 125–144 (2010)

Computer Modelling in Physics Education: Dealing with Complexity

Onne van Buuren⬚ and André Heck⬚

Abstract Computer modelling and analysis of measurements with ICT enable secondary students to study situations that are both more realistic and more complex. To fully benefit from these advantages, students need to grasp the structure of the physics domain under study. In a design research project, three ways to help the students grasp the structure of one-dimensional Newtonian dynamics were included into the lesson materials for grade 10 students, namely, (1) a map of this physics domain, (2) goal-free exercises, and (3) graphical models. Students were asked about their opinions on the effectivity of these three ways. The students considered goal-free exercises and graphical models effective for grasping this structure. They found the map rather complicated and only effective when carefully introduced.

Keywords Modelling · System dynamics · Graphical modelling · Experimenting with ICT · Cognitive load

Introduction

An important advantage of computer modelling and analysis of measurements with ICT in secondary education is that the computer can do calculations that otherwise are too cumbersome or too complicated. With appropriate tools, students can study situations in which (1) more quantities are involved, (2) more and (3) more difficult relations between quantities take part, and (4) more than only one or two quantities are variable objects, that is, have values that change during the physical process under study. In other words, with ICT, students can study situations of greater complexity.

O. van Buuren (✉)
Haags Montessori Lyceum, The Hague, The Netherlands
e-mail: OnnevanBuuren@quicknet.nl; o.p.m.van.buuren@vu.nl

O. van Buuren · A. Heck
Korteweg-de Vries Institute for Mathematics, University of Amsterdam, Amsterdam, The Netherlands
e-mail: a.j.p.heck@uva.nl

© Springer Nature Switzerland AG 2019
E. McLoughlin and P. van Kampen (eds.), *Concepts, Strategies and Models to Enhance Physics Teaching and Learning*,
https://doi.org/10.1007/978-3-030-18137-6_8

This has several consequences for physics education. First, as an advantage, the situations to be studied can be more realistic than the examples commonly found in school textbooks. Furthermore, the increased level of realism facilitates the comparison of model and experiment. In general, it is easier to do measurements in a more realistic situation than to create apt experimental conditions for a simplified situation. Because complicating factors can be taken into account in the computer model, there is less need to eliminate these factors in the experimental situation. For example, in a study of the movement of a falling body, one does not need to eliminate air friction in the experimental situation because, in many cases, air friction can be easily accounted for in a computer model.

A second consequence is that students can study an entire process, and they can study it from many points of view, because with ICT, complete graphs of all quantities involved can easily be generated. In this way, students are not limited to doing exercises in which only constant, average, or instantaneous values of a limited number of quantities must be calculated. They can focus on larger structures, formed by networks of relations within the model. This certainly holds for graphical modelling, that is, when a graphical system dynamics based modelling tool is used. According to Van Borkulo [1], graphical modelling is indeed advantageous for studying relatively complex situations. Students may even grasp the main, generative structure of the field of interest. For example, in case of one-dimensional Newtonian dynamics, this structure consists of the equations of motion. These equations are shown in Fig. 1, together with the corresponding graphical model. When students grasp this structure, their main task is to determine the net force F_{net} for each specific situation and hereafter study the situation with the computer model.

But there are drawbacks. Although computer modelling and ICT-supported experimentation enable the study of more complicated physics, they also increase the cognitive load for the learners because many competencies are required at the same time. Both computer modelling and the use of ICT for measurements take considerable time to learn, so students better start with it at an early age. Van Buuren [2] developed a learning path on (graphical) modelling in combination with ICT-supported experimentation for the first two years of lower secondary physics education (grades 8 and 9 in the Netherlands). At the end of this learning path, students could understand the

Fig. 1 Graphical model of the main structure of one-dimensional Newtonian dynamics

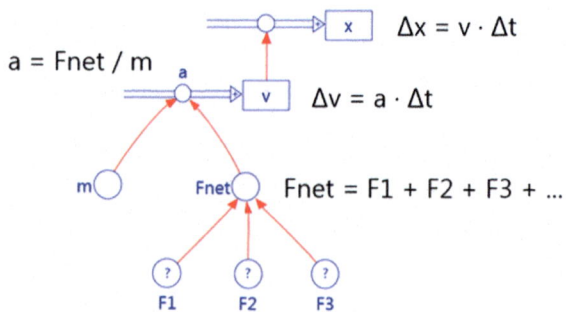

structures of graphical models and they were able to construct very simple models, consisting of two relations.

Another drawback was encountered in a pilot extension of the aforementioned learning path on dynamics in grade 10. Students had difficulty in dealing with the greater number of choices and the more extended structures in the new tasks. In interviews, students mentioned several reasons for these difficulties. First, they told that they tend to focus on calculations instead of on structures. One reason is that they are used to doing so. Another reason is that the students expect the focus in examinations to be on calculations, not on modelling or experimenting. Second, students reported that they were not used to tasks in which more than only a few steps had to be made. If a task could not be accomplished that easily, frustration set in. Finally, students noticed that they had not yet learned to divide problems into small, useful steps themselves. They stated that they had not sufficiently grasped the structure of the physics to do so. The greater number of choices and the more extended structures they had to deal with apparently contributed to a substantial increase of cognitive load. Of course, it is possible to diminish this cognitive load by letting students only study less complicated models and perform only minor analyses of measurements, but then one would not take full advantage of the benefits of ICT. Therefore, the question is how secondary school students can acquire a sufficiently comprehensive view of a physics domain in order to be able to learn from computer modelling and from ICT-supported analysis of measurements.

To answer this question, we redesigned the module for grade 10 with the objective of helping the students to grasp the structure of one-dimensional Newtonian dynamics. We tested this redesigned module in class and asked the students via a questionnaire what they thought of the effectiveness of the three most important measures that we took to help students grasp this structure. In this paper, we discuss these three measures and the results of the questionnaire.

Design Considerations

In the redesigned module, we tried to help students grasp the structure of one-dimensional Newtonian dynamics in three different ways:

- by offering a map of one-dimensional Newtonian dynamics;
- by adding so-called goal-free exercises to the module;
- by using graphical models.

These three ways have one thing in common: they all aim at the construction of schemata. Work on novice-expert differences suggests that access to schemata in memory is a critical characteristic of skilled performance (see, for example, [3]). Therefore, an important function of learning is to store automated schemata in long-term memory. Below, these three ways are briefly discussed.

A Map of One-Dimensional Newtonian Dynamics

The idea of offering students a map of one-dimensional Newtonian dynamics stems from the pilot experiment mentioned in the introduction. Some students asked for help, individually or in small groups, because they felt that they did not understand one-dimensional dynamics well enough. As a scaffold, we constructed and simultaneously explained this map to these students. It was much appreciated. The map consists of two main routes through Newtonian dynamics, a so-called 'experiment route' and a 'modelling route'. These routes are depicted in Fig. 2.

The experiment route typically starts with results obtained by measuring positions. By taking derivatives with respect to time, velocities and, subsequently, accelerations can be determined. Typical techniques are the use of tangents for determining slopes and calculating average velocities and average accelerations over intervals of time. Finally, by using Newton's second law, information is gained about the forces acting on the moving object.

The modelling route typically starts with establishing the forces on the object and subsequently determining the net forces and the accelerations. By means of

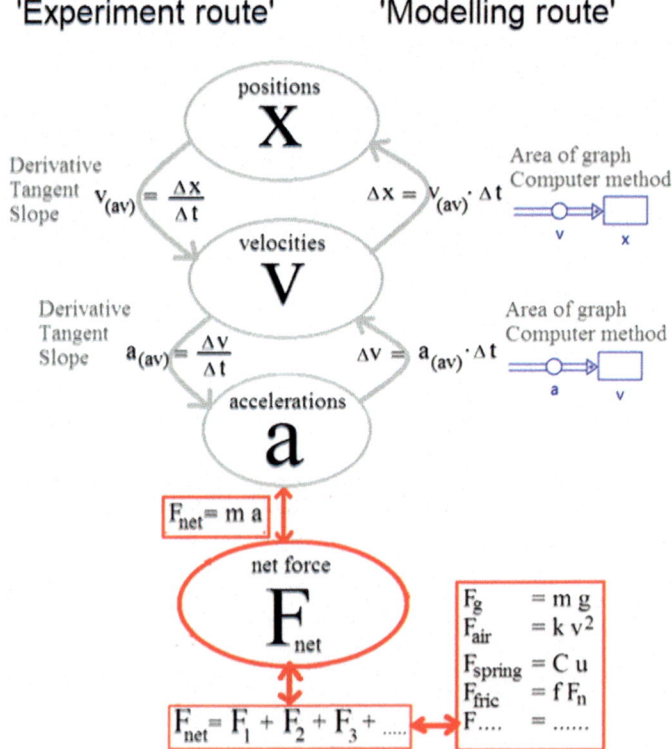

Fig. 2 A map of one-dimensional Newtonian dynamics

numerical or analytical integration, velocities and, subsequently, positions are determined. Graphical models can be positioned on this route. We added this map to the redesigned textbook and incorporated a few exercises in which students could become familiar with this map.

Goal-Free Exercises

The idea of using goal-free exercises stems from cognitive load theory (CLT). In CLT, a distinction is made between means-ends and goal-free exercises [3]. In a means-ends exercise, a student is asked to find a specific answer to a question, e.g., given a set of data, one or only a few specific quantities must be found. A student faced with such a problem is unlikely to have a schema to generate a solution, and is likely to try to solve a problem through means-ends analysis. Working backward from the goal, the student searches for a series of adequate equations connecting the goal to the givens. When such a series is found, (s)he can work forward calculating values for the unknown quantities. For an expert, this can be an efficient problem-solving approach. But, as Sweller [4] found, for a novice who lacks automated schemas for problem solving it is a process that imposes a heavy extraneous cognitive load on working memory capacity which bears little relation to schema construction.

A goal-free exercise can consist of the same data in the same situation, but the learner is not asked for a particular quantity. Instead, (s)he is asked to find as many quantities as possible, or, as we did, to explain what can be found and how this can be done, using the given data. In this way, the student is actually asked to express in their own words what is going on in physics terms and is in this way stimulated to create a schema for the situation, in which not only a few specific quantities, but all quantities are connected. Although the result may be identical (the same quantities can be found), according to Sweller [4] the strategy used and the cognitive processes that occur are quite different. In this goal-free approach, cognitive load is greatly reduced and the learner is focused on schema construction instead of finding one specific answer. An example of a goal free exercise is shown in Fig. 3. About 10% of all exercises in the redesigned textbook are goal-free.

Graphical Models and Graphical Modelling

In Fig. 1, an example of a graphical model is shown; it is an incomplete model that represents the main structure of one-dimensional dynamics. In a graphical model, the variables and relationships between variables are visually represented as a system of icons in a diagram. The graphical equivalent of a difference equation is the combination of a 'stock variable', represented by a rectangle, and one or more 'flow variables', represented as 'thick' arrows. Variables that are not explicitly part of a difference equation are referred to as 'auxiliary variables' and are represented by

Exercise 3.17: breaking car

Explain what can be found and how this can be done with the data in this graph.

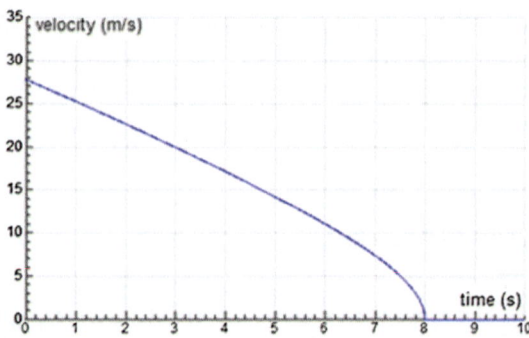

Fig. 3 Typical example of a goal-free exercise. The student is not asked to determine a specific quantity, such as the distance travelled or the acceleration at a specific moment. Instead, (s)he is asked to explain what results can be obtained from the given data

circular icons. Flow variables and auxiliary variables are defined by direct relations. By a direct relation, we mean a mathematical relationship between symbolized quantities in which at least one quantity can be isolated and written as a closed form expression of the other quantities. Contrary to difference equations, direct relations must be entered explicitly (as formulas) into the computer environment. In the graphical diagram, they are not immediately visible, but 'connectors' ('thin' arrows) show their presence.

Forrester [5] developed these graphical diagrams as an intermediate transition between a verbal description and a set of equations. Their main goal is to communicate the causal assumptions and the main features of the model. Several researchers (see, for example, [6]) have suggested that the visual representations in graphical models can help to shift the focus from learning and working with mathematical formulas to more qualitative conceptual reasoning. From the CLT perspective [3, 7–9], success of graphical modelling can be attributed to the assumption that the use of a graphical external representation supports the off-loading of working-memory and allows the freed working memory to be used for learning. The links between the elements of the diagrams and their mathematical counterparts are deliberately not visualized for the purpose of allowing modelers to focus on the main structures of the model. However, as Groesser [10] points out, graphical models are only helpful if the modeler is familiar with the graphical modelling language. By not presenting the graphical elements and mathematical counterparts simultaneously, for novice modelers, extraneous load is in fact increased.

Materials and Methods

In this section we present the design of the lesson materials, the setting in which the research took place, and the research instruments.

Design of the Lesson Materials

Two modules on one-dimensional Newtonian dynamics for grade 10 have been designed, one for vwo, the Dutch pre-university stream, and one for havo, the Dutch general upper education stream. At the start of grade 10, there are only minor differences between these streams; therefore, the two modules are quite similar. Activities have been implemented in the computer learning environment COACH [11] because COACH can be used both for modelling and for doing and analyzing measurements, and because it is available at the majority of Dutch secondary schools. In each module, as many known conceptual problems of students as possible are dealt with. Modelling activities and experiments are not just add-ons; they are fully integrated into the lesson materials.

The module starts with a repetition of topics from grade 9: movement (including the usage of tangents in position-time graphs and the meaning of area in velocity-time graphs), forces, and Newton's first and second law. Subsequently, explicit attention is given to systematic problem-solving strategies and to the construction of force diagrams, because constructing these 'force models' can be considered as the kernel of modelling in dynamics. In a next section of the module, the map of one-dimensional dynamics is introduced. Hereafter, without computer, mathematically simple cases are modelled, such as limiting cases and cases of equilibrium. These simple cases are useful later on, when more complex models must be evaluated. To prepare students for a complete analysis of video measurements, in a separate section, advanced techniques for analyzing graphs with COACH are introduced: filtering, function-fitting, using tangents and determining derivatives. Finally, graphical models are reintroduced, following a relation approach. In this approach, the relations between the quantities are considered as the fundamental building blocks of the model [12]. The module ends with modelling activities, in which students compare models for realistic situations with results of experiments. A typical example of such an activity is described in [13]. In the module, students must construct parts of models or extend existing models, they are not yet asked to construct a complicated model from scratch.

Research Setting

Classroom experiments were carried out in two vwo-4 classes and in one havo-4 class (all grade 10 students of age 15–16 year) at a Montessori secondary school. The course took 37 h of lesson-time. Students' progress was monitored with several regular assessments. Two physics teachers were involved. Unfortunately, in the previous school year, many of the students for whom the module has been designed had not finished the part of the physics course on dynamics and graphical modelling. This put much more demands on the part of the module that recapitulates the expected knowledge about dynamics and graphical models than would otherwise have been necessary. In addition, many students did not perform strongly in physics nor in many other subjects. For example, at the end of the year, the average mark for mathematics, chemistry, and physics of 40% of the participants did not meet standards. This makes it difficult, if not impossible, to compare results with findings from other experimental studies, such as the pilot experiment mentioned earlier in this paper. Furthermore, the teachers involved paid little attention to the preparation of the lessons; they expected that their students could work with the lesson materials independently. For example, they paid no attention to the map of one-dimensional Newtonian dynamics and its purpose for the lessons. This has certainly influenced the outcomes of the study.

Research Instruments

Our main research instrument was the questionnaire in which we ask students to express their opinions on the effectiveness of the three ways in which we tried to help them grasp the structure of Newtonian dynamics. When students' answers in the questionnaire were unclear, we asked them in small-scale audio recorded in-depth interviews to clarify these answers. In addition, we made classroom observations. Finally, as a measure for the general science performance level of the students, we collected their final marks at the end of the school year and took the average of the final marks for mathematics, chemistry, and physics.

The questionnaire was given to the vwo-4 students a few weeks after their final assessment of dynamics. In havo-4, there were four months between the final assessment and the questionnaire. To refresh the students' memories, we added a booklet to the questionnaire in which examples are shown of relevant elements of the educational materials, such as the map of dynamics, a few goal-free exercises and a graphical diagram. The content of the questionnaire is the following.

Question 1 intends to measure the students' general feelings about their grasp of the structure of dynamics: "Do you feel, at this moment, after the final assessment, that you have a good grasp of the structure of dynamics?".

Question 2 concerns to what extent students consider modelling and doing experiments helpful for developing a grasp of the structure compared to other educational activities. The students are given the list of Fig. 4 and are asked which of these activities they consider helpful in this regard.

Question 3–6 intend to measure the students' opinions on the effectiveness of the map of one-dimensional Newtonian dynamics. The students are asked whether they remember this map, whether they have used it, and if it is helpful. In the last question of this series, the students are asked to explain their answers.

Question 7–9 intend to measure the effectiveness of goal-free exercises. Students are asked whether they remember such exercises and whether they are helpful grasping the structure. Finally, students are asked to explain their answers.

The last two questions, numbered 10 and 11, are about graphical models. Because graphical models only can be helpful if they are understood, students are first asked to what extent they understand graphical models. Subsequently, they are asked whether

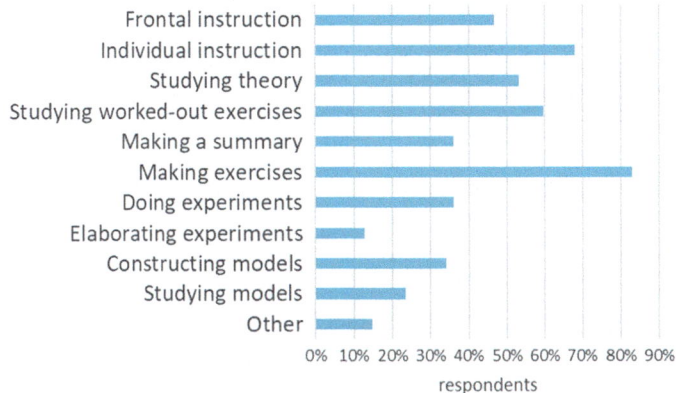

Fig. 4 Students' answers to the question what helps to grasp the structure of dynamics

they think that graphical models can help to grasp the structure of the situation the model is about. Again, they are asked to explain their answers.

Findings

Students' Answers to the General Questions on Structure

Of all 53 students, 47 (89%) handed in a completed questionnaire. Of these 47 students, 57% stated they had somewhat grasped the structure of Newtonian dynamics, 23% stated they had grasped it well, and the remaining 20% declared that they hardly had grasped it. All but one of the latter students performed poorly in science and mathematics in general. Apparently, there is some correlation between students feeling they have grasped the structure and their general performance in science and mathematics.

As can be seen in Fig. 4, traditional methods of education are clearly favored by the students over doing experiments or constructing models, although there may be a bias towards items in the upper part of Fig. 4. Making exercises is considered helpful by the most students (83% of all respondents), whereas doing experiments is considered useful by only 36% and constructing models by only 34% of the respondents. For two reasons, this is not surprising. First, students tend to consider exercises as more important as a training for assessments. Second, before students can appreciate the larger structures that come to the fore in experiments and models, they must learn to connect quantities to each other on the smaller scale of exercises. From this point of view, it is interesting that constructing models is appreciated as a means for grasping the structure by 73% of all the students who in question 1 stated that they had grasped it well. We did not find a similar effect with respect to doing experiments. It appears that grasping the structure well correlates with an appreciation of constructing models, but not with an appreciation of doing experiments. Finally, we note that constructing models is appreciated more by students than studying given models.

Students' Opinions on the Map of One-Dimensional Dynamics

Practically all respondents could remember the map in Fig. 2, but most students had not used it at all (53%) or only a few times (27%). None of the students had used it often. Only 15% of all students stated that the map was really helpful, a small majority (53%) stated that it helped a bit. Of all respondents, 28% stated that it did not help at all. Nine percent of the respondents commented that the map is a useful summary, 15% stated that it helps to grasp the structure and helps determining which steps must be made in several situations. Half of all students found the map complicated, disordered, unclear, or even discouraging at first glance ("Wow, this must be difficult!"), but several of these students added that the map is helpful when it is properly understood: it should have been explained by the teacher. A few students gave another important argument: "It is complicated, because you already must understand all individual formulas". A few other students found the map superfluous: they already had sufficiently grasped the structure.

We conclude that the map can be helpful, but only provided that it is constructed together with the students and explained by the teacher, and that students already understand the individual elements of the map.

Students' Opinions on Goal-Free Exercises

Of all respondents, 91% remembered the goal-free exercises. These exercises helped very well with grasping the structure according to 47% of the respondents. Another 43% found that they helped, but not better than other exercises. Typical comments of students in favor of goal-free exercises are the following:

- You gain more insight into what you can do with the information.
- They show what it is all about.
- You need to use all your knowledge.
- They help to learn to make more than one step.
- You are more actively involved.

These comments fit well to the idea of schema construction. The few clear student arguments against goal-free exercises are that some students get confused, do not know what to focus on, or easily forget things. Apparently, the lack of a specific goal or the seemingly boundless character of the answer can make a minor part of the students feel uncertain, afraid to forget things. Finally, a few students reported that these exercises can only be effective with sufficient basic knowledge.

We conclude that students consider goal-free exercises as helpful for grasping the structure.

Students' Opinions on Graphical Models

Of all respondents, 59% declared that they understood graphical models well, 27% stated that they understood them reasonably, but not very well, and 17% found that they understood them only a little, or not at all. On the question whether graphical models, according to the student, can help to grasp the structure of what is modelled, 74% of the respondents answered in the affirmative, 11% answered negatively, and 15% gave no or no clear answer. Of all respondents, 43% explicitly commented that the graphical diagrams clearly show the relations between variables or, in more general words, that the diagrams help grasp the structure. Two other students stated that the diagrams help to approach the situation in a more stepwise manner. The advantages are not unconditional: 13% of the respondents remarked that the graphical model must be understood first and needs explanation, and 6% of the respondents noted that the graphical model must not be too extended. Only a few students preferred formulas or other knowledge above graphical diagrams. Some students explicitly stated that constructing graphical models is more effective than just using or studying them.

Influence of Performance Level

Unless stated otherwise, we did not find clear correlations between general performance level in science and mathematics of the students and their opinions on the effectivity of the ways for grasping the structure. We did not make a thorough statistical analysis, however; that is beyond the scope of this chapter.

Conclusions and Discussion

After the course, most students felt they had at least somewhat grasped the structure of one-dimensional Newtonian dynamics. Students who felt otherwise performed poorly in mathematics and science in general. For gaining insight into the structure, traditional methods of education, and especially making exercises, are favored over doing experiments or constructing models. One reason may be that students consider exercises as the most important as a training for assessments, but another reason appears from the students' comments. Before students can appreciate the larger structures that come to the fore in experiments and models, they must learn to connect quantities to each other on the smaller scale of exercises. A basic understanding of the individual elements is a prerequisite for grasping the structure, and exercises and instruction are ways to build up this basic understanding. This is in agreement with findings of Van Borkulo [1] that graphical modelling may not be advantageous for the learning of simple conceptual domain knowledge. It also explains why those students who after the course stated that they had grasped the structure well, appreciated constructing models better as a means to achieve this.

Among the exercises, in accordance with cognitive load theory, goal-free exercises are considered by the students as especially helpful for grasping the structure. A disadvantage is that the open character of goal-free exercises can make some students feel uncertain. An advantage is that goal-free exercises can be used in an early stage of the course.

The map of one-dimensional Newtonian dynamics was not successful as a means to develop a grasp of the structure. The main cause is that the map is complicated, having a high cognitive load which should have been diminished by an introduction by the teachers, but it was not introduced. A second cause is that probably not all students mastered the individual elements of the map at the moment it was presented in the lesson materials. Students who participated in the earlier pilot project were definitely more positive about the map. We conclude that the map can be helpful, but only provided that the two causes mentioned above are sufficiently dealt with.

Most students stated that they understand graphical models at least reasonably well. Three quarters of the respondents found that they help grasp the structure of the situation. In their comments, many students mentioned that the graphical models clarify relations in the model. These findings are in accordance with the goal of the graphical diagrams as intended by Forrester [5], namely, to communicate the causal assumptions and the main features of the model. The comments of part of the students support the statement by Groesser [10], that graphical models are only helpful if the modeler is familiar with the graphical modelling language. Anyway, our students clearly favor the graphical diagrams over the map of one-dimensional dynamics, in which too many elements are presented simultaneously. Here, we note that the map contains more information than a graphical model because it also contains the experiment route through dynamics.

The question whether we indeed succeeded in helping students grasp the structure of Newtonian dynamics cannot be answered through this study. We need to develop reliable instruments that measure that directly, instead of opinions. But our results indicate that graphical modelling indeed contributes to the development of a grasp of the structure.

Acknowledgements This research was financed by the Dutch Ministry of Education, Culture, and Science, as part of the Postdoc-VO project.

References

1. Van Borkulo, S.: The assessment of learning outcomes of computer modeling in secondary science education. Doctoral Dissertation, Twente University. http://ris.utwente.nl/ws/portalfiles/portal/6084208 (2009). Last accessed 26 Nov 2018
2. Van Buuren, O.: Development of a modelling learning path. Doctoral Dissertation, University of Amsterdam. http://hdl.handle.net/11245/1.416568 (2014). Last accessed 04 Dec 2017
3. Sweller, J., Van Merriënboer, J., Paas, F.: Cognitive architecture and instructional design. Educ. Psychol. Rev. **10**(3), 251–296 (1998). https://doi.org/10.1023/A:1022193728205
4. Sweller, J.: Cognitive load during problem solving: effects on learning. Cogn. Sci. **12**(2), 257–285 (1988). https://doi.org/10.1207/s15516709cog1202_4
5. Forrester, J.: Industrial Dynamics. MIT Press, Cambridge, MA (1961)
6. Niederrer, H., Schecker, H., Bethge, T.: The role of computer-aided modelling in learning physics. J. Comput. Assist. Learn. **7**(2), 84–95 (1991). https://doi.org/10.1111/j.1365-2729.1991.tb00231.x
7. Sweller, J.: Cognitive load theory, learning difficulty, and instructional design. Learn. Instr. **4**(4), 295–312 (1994). https://doi.org/10.1016/0959-4752(94)90003-5
8. Van Merriënboer, J., Sweller, J.: Cognitive load theory and complex learning: recent developments and future directions. Educ. Psychol. Rev. **17**(2), 147–177 (2005). https://doi.org/10.1007/s10648-005-3951-0
9. Sweller, J., Ayres, P., Kalyuga, S.: Cognitive Load Theory. Springer, New York, NY (2011)
10. Groesser, S.: Model-based learning with system dynamics. In: Seel, N. (ed.) Encyclopedia of the Sciences of Learning, pp. 2303–2307. Springer, New York, NY (2012)
11. Heck, A., Kedzierska, E., Ellermeijer, T.: Design and implementation of an integrated computer working environment for doing mathematics and science. J. Comput. Math. Sci. Teach. **28**(2), 147–161 (2009)
12. Van Buuren, O., Heck, A., Ellermeijer, T.: Understanding of relation structures of graphical models by lower secondary students. Res. Sci. Educ. **46**(5), 633–666 (2015). https://doi.org/10.1007/s11165-015-9474-x
13. Heck, A., Van Buuren, O.: Ramp it up and down. Phys. Educ. **52**(1), 015014 (2017). https://doi.org/10.1088/1361-6552/52/1/015014

Development of Data Processing Skills of Physics Students in Intermediate Laboratory Courses

Inkeri Kontro ⓘ

Abstract Laboratory courses are an essential part of physics education, and the need for learning goals that focus on laboratory skills rather than physics contents has recently been emphasized. We evaluated the effect of skills-oriented laboratory courses on data processing skills with the Concise Data Processing Assessment (CDPA). Though no overall improvement occurred, changes in answering patterns in CDPA occurred. The students showed favourable shifts in their answer patterns for items dealing with fitting errors but small declines in identifying power laws in data. This is likely due to course curriculum, which emphasized fitting. Although the sample size is small, the results indicate that CDPA answers can provide information on learning of specific topics even if changes in the overall score are minimal. The measured data processing skills did not correlate with expert-like attitudes (as measured by the Colorado Learning Attitudes about Science Survey for Experimental Physics) or course grades.

Keywords Undergraduate · Laboratory · Student attitudes · Data processing

Introduction

Laboratory work is an essential part of most university physics education. However, traditional laboratory work does not foster conceptual understanding [1] and laboratory work is most useful in teaching students skills rather than physical concepts, which is why the American Association of Physics Teachers recommends assigning skills-based learning goals to laboratory courses [2]. Two important skills to learn are data processing and error estimation. The ability to understand the reliability of data is a fundamental skill for any scientist, and for physicists, a natural surrounding for learning this skill is the laboratory where the data are produced. However, there are few standardized tests for measuring data processing skills. One such test is the

I. Kontro (✉)
Department of Physics, University of Helsinki, POB 64, 00014 Helsinki, Finland
e-mail: inkeri.kontro@helsinki.fi

© Springer Nature Switzerland AG 2019
E. McLoughlin and P. van Kampen (eds.), *Concepts, Strategies and Models to Enhance Physics Teaching and Learning*,
https://doi.org/10.1007/978-3-030-18137-6_9

Concise Data Processing Assessment (CDPA) [3], which is a multiple-choice test that consists of ten questions. In CDPA, first-year students typically score points corresponding to pure guess (23.5%) and faculty score on average 80% [3].

In addition to skills, interest in the PER community has recently focused on the development of expert-like attitudes during laboratory courses. Use of the Colorado Learning Attitudes about Science Survey for Experimental Physics (E-CLASS) [4] has shown that generally, traditional laboratory courses have a detrimental effect on the expert-like attitudes, whereas courses that focus on skills see small improvements [5].

The intermediate (2nd–3rd year) laboratory courses at the University of Helsinki are a pair of half-semester long courses, which span the spring term. The courses have been reformed with a focus on learning goals [6]. The reform led to a high attainment of learning goals. While the students attained the learning goals of the courses, expert-like attitudes (as measured by E-CLASS) remained unchanged [6]. The teachers and students were satisfied with the course, but to study the relationship between learning, student attitudes and student experience further, we administered the CDPA survey and the E-CLASS survey, and compared the results to course grades. The goal was to see whether favourable attitudes towards experimental physics correlate with data processing skills before or after instruction. Their correlation with the grades were also measured. The grade summarizes the learning goals of the course, which includes but is not limited to data analysis skills and expert-like practices in the laboratory. The research questions are thus whether the standardized instruments measure learning in this course, and whether the course learning goals reflect all of the aspects to which we aspire.

Data Collection and Analysis

Students completed problem solving exercises for course credit. The responses to the surveys in this study were collected as part of these exercises, and credit corresponding to a single problem was awarded for participation. Only the responses of the participants who gave consent for use of their responses in research are included in the analysis. The response rate for each questionnaire was high, 70–90%. Twenty-six students answered both CDPA questionnaires, and 22 students answered both CDPA questionnaires and the E-CLASS pre-test.

The CDPA survey was administered during the first week of teaching and after the mid-term break, at the beginning of the second course, when the relevant material had been covered. The E-CLASS surveys were administered during the second week of the first course and at the end of the second course. The surveys were spaced out to reduce the burden on a single week.

The CDPA was scored simple awarding 1 point for each correct answer. The normalized gain g of CDPA was calculated from

$$g = \frac{post - pre}{100 - pre}, \quad (1)$$

where post and pre are the scores of the post- and pre-test, respectively [7].

The E-CLASS survey was asked both with the formulation "What do you do when performing an experiment" and "What would an experimental physicist say". The E-CLASS survey was marked by collapsing the 5-point Likert scale (strongly agree − strongly disagree) into categories of agree, disagree and neutral with respect to the expert answer, and assigning 1, −1 and 0 points for these response types, respectively [4]. In addition to the survey results, the data set consists of students' grades from the first course. These grades are overall grades for the course; grades for individual learning goals were not available. Grades were administered on a scale of 0–5, where 0 is fail, 1 the lowest passing grade and 5 the highest grade. No students achieved grades of 0–1 on this course.

Results and Discussion

Student Understanding of Error

The average CDPA score on the intermediate laboratory course was (4.04 ± 0.41) in the post-test. Day and Bonn reported values of 3.03 and 4.73 for second and fourth year students, respectively [3]. The students on the intermediate laboratory course are generally second- to third year students. A few may be older, as physics students do not necessarily complete the bachelor degree in three years and physics minors may take the course as part of their master's degree.

The CDPA score showed no change with instruction. The results from the pre- and post-tests are shown in Table 1. The item difficulty varied significantly, and largely followed that presented in [3]: items 1, 5 and 6 were the most difficult. The only major deviation from this was the difficulty of item 4, in which the University of Helsinki students did more poorly than expected. This item concerns a log-log plot with a power law.

While no change in the overall score emerged, some answering patterns warrant a closer look. The two most difficult questions, items 5 and 6, concern linear fits of data with error bars of varying size. The improvement for the correct choice was modest for question 5 and zero for question 6, but there was a marked shift in the answers (Table 2). The right answer, naturally, was to fit the line to best match the data points with the smallest error bars (choices 5b and 6a, respectively). The answer choices were as follows (for question 5)

- (a) weighing of more uncertain data points
- (b) (correct) weighing of more precise data points

Table 1 CDPA scores (N = 26), average score

Item	Pre	Post	Gain	g
Overall	3.92 ± 0.35	4.04 ± 0.41	0.12 ± 0.35	−0.01 ± 0.06
Question 1	0.23	0.31	0.08	
Question 2	0.77	0.77	0.00	
Question 3	0.31	0.35	0.04	
Question 4	0.27	0.31	0.04	
Question 5	0.08	0.19	0.11	
Question 6	0.12	0.12	0.00	
Question 7	0.62	0.46	−0.16	
Question 8	0.58	0.46	−0.12	
Question 9	0.77	0.69	−0.08	
Question 10	0.35	0.35	0.00	

Overall score with standard error of the mean and scores for individual questions pre- and post instruction, gain and normalized gain g

Table 2 CDPA answers for items 5 and 6 ($N = 26$)

Question 5	a	b	c	d
Pre-test	0	2	17	7
Post-test	1	5	6	14
Question 6	a	b	c	d
Pre-test	3	1	6	16
Post-test	3	3	11	9

The right answer is underlined and the wrong answer, which still considers error bars (for questions 5 and 6), is in italics

- (c) unweighted fit
- (d) intermediate fit weighing more precise data points somewhat more than the uncertain ones

For question 6, the answer choices were

- (a) (correct) weighing the most precise data points
- (b) unweighted fit
- (c) touching the most error bars
- (d) fitting an equal amount of data points below and above the line

Prior to instruction the most popular choice for question 5 was the unweighted fit (5c, 17 answers), with the moderately weighted choice (5d, 7) being second. Post instruction, the percentages were almost reversed (6 and 14 answers for 5c and 5d, respectively). In addition, the amount of correct answers had increased slightly (Table 2).

For question 6, the pre-instruction favourite was equal amount of points over and under the fit (6d, N = 16) with the fit touching most error bars being second

Table 3 CDPA answers for items 7 and 8 ($N = 26$)

Question 7	$a*$	b	$\underline{c}$	$d*$
Pre-test	4	5	16	1
Post-test	3	9	12	2
Question 8	$\underline{a}$	$b*$	$c*$	d
Pre-test	15	3	4	4
Post-test	12	6	3	5

The right answer is underlined and the answers containing exponentials are in italics. Answers with incorrect units are marked with asterisks

Table 4 CDPA answers for item 9 (N = 26)

Question 9	$\underline{a}$	b	c	d	e
Pre-test	20	3	3	0	0
Post-test	18	1	7	0	0

The right answer is underlined and the answer that considers measurement error larger than rounding error is in italics

(6c, N = 6), and the correct option third (6a, N = 3). The prevalence of 6c increased in the post-test, though the changes in the answer distribution are less marked than for item 5.

The changes in the answer patterns in particular to question 5 suggest that while the students have not fully grasped the extent to which the differences in error of data points affect fitting, they are aware that the error bars play a role. Answers that consider error bars were more common after instruction.

However, is it fair to interpret these shifts favourably, when there are questions with non-favourable changes? The questions with the largest losses in right answer concerns identifying a power law graph (question 7) and identifying a power law in a table (question 8). In both questions, the incorrect assumption of exponential relationship is somewhat more common in the post-test. The other wrong choices have the correct power law with incorrect units (7d, 8b and 8c). The answer distributions to these questions are shown in Table 3.

Power laws do not feature in the intermediate laboratory courses, however, exponentials and (semi)logarithmic plots do. Post-instruction, students opted slightly more for exponentials, which shows they may have opted for familiarity. In addition, the skill to recognise power laws is not explicitly taught in the laboratory courses or any previous courses (Szabolcs Galambosi, private communication) which may explain both the declining results of items 7 and 8 and the surprising difficulty of item 4, which also concerns a power law.

The third question with a negative gain, item 9, concerns rounding and measurement error of a scale measuring the mass of several objects, when the mass of one was known. Here, students were more prone to overestimating the error in the post-test by conflating the rounding error and the resolution of a scale (Table 4). The answer choices to question 9 were, in order:

(a) (correct) propagation of rounding error
(b) conflating the uncertainty of the mass of one object with the resolution of the
     scale
(c) conflating the rounding error of a single measurement with the resolution of the
     instrument (considered by some experts to be correct)
(d) ignoring uncertainty entirely
(e) not treating the rounding uncertainty symmetrically.

Hence, some students changed their answer from correct to overestimating the error, according to the CDPA scoring. However, some experts also consider this overestimation (9c) the correct choice [3], as the error of a measurement device is usually larger than just the rounding error. Indeed, during the laboratory course, the students familiarized themselves with instrument manuals that consistently state that the instrumental error is a function of, and larger than the resolution. Scales did generally not feature the students' design of lab experiments, but the uncertainty of e.g. multimeters had to be considered in error calculations.

The conclusion is that there are small signals that the instruction affected the students' answer both for better and for worse. In linear fits, the students were more likely to choose answers that considered error bars, whether or not to the necessary degree. However, non-exponential relationships were slightly increasingly identified as exponential. Of course, with such a small sample, in particular the smaller changes in answers are likely to be the product of chance. The large change in the answering pattern to item 5 likely indicates that students learned to consider the error bars during the course, even if they were not able to correctly interpret the necessary weighing.

Correlation of Expert-like Attitude, Experience and Skill

The correlations between E-CLASS (pre-test) and CDPA scores is presented in Table 5. The number of complete datasets dropped to 22, and dropped further, when the E-CLASS post-test was added. For this reason, the E-CLASS post-test is not considered here, as the reliability of any results would decline further.

Curiously, there seems to be no correlation between CDPA scores and the student's expert-like attitudes, nor their perception of expert attitudes as measured by the E-CLASS questionnaire. CDPA and E-CLASS results did also not correlate with the students' grades from the laboratory course (Table 5). This is surprising, as we expected students who understand error analysis better to achieve higher grades. On the other hand, no very low-achieving students attended the course, which may skew the results.

The grading rubrics on the course were based on learning goals and explicitly covered data processing. The absence of correlation between CDPA scores and grades may be due to obfuscation by the other learning goals, such as experiment design, report writing and programming skills. As previously discussed, the CDPA covers a variety of topics that did not feature on the course. The students on these courses

Table 5 The correlation coefficient r between CDPA pre-instruction and post-instruction scores, CDPA normalized gain g, E-CLASS scores for questions phrased "What do you think when doing labs" (E-CLASS own) and "What would an experimental physicist say" (E-CLASS phys) and grade ($N = 22$)

	CDPA pre	CPDA post	CDPA g	E-CLASS own	E-CLASS phys	Grade
CDPA pre	1					
CPDA post	0.50	1				
CDPA g	−0.37	0.55	1			
ECLASS own	0.04	0.10	0.10	1		
ECLASS phys	−0.10	0.24	0.27	0.45	1	
Grade	−0.03	0.04	0.09	0.27	0.21	1

generally reached the established learning goals and the students considered the course useful and challenging [6].

The low correlation between CDPA normalized gain and post-test scores ($r = 0.55$, adjusted $r^2 = 0.27$) is statistically significant ($p < 0.01$) but simply explained by that the students who improved their score during instruction were more likely to do better on the post-test. The very modest negative correlation between CDPA pre-test score and gain, which could indicate that students were more likely to learn error analysis if their prior knowledge was higher, was not significant ($p > 0.05$).

Conclusions

Classifying and identifying the CDPA answers indicates that students on the intermediate laboratory courses paid more attention to the effect of measurement error in linear fits, a topic that featured heavily in the exercises on the courses. The students found questions involving power laws more difficult than expected. An examination of the curriculum shows that they are rarely exposed to data for which they need to find a suitable model without constraints, and power laws do not feature on courses except in specific contexts. This warrants a critical look at the curriculum. Expert-like attitudes did not correlate with data processing skills or achieving the general learning goals of the course, as measured by student grades.

Overall, CDPA seems to provide useful information on specific improvements and problems in the students' knowledge even when changes in overall score are not present. As such, it is a useful tool for probing not only the level of student knowledge but common misconceptions.

Acknowledgments The author thanks Dr. Szabolcs Galambosi and Olga Heino for discussion of the results, and O. H. and Dr. Tommi Kokkonen for the discussion of the translation of the CDPA. The financial support of the Waldemar von Frenckell foundation is gratefully acknowledged.

References

1. Royuk, B., Brooks, D.W.: Cookbook procedures in MBL physics exercises. J. Sci. Educ. Technol. **12**(3), 317–324 (2003). https://doi.org/10.1023/A:1025041208915
2. MacIsaac, D.: Report: AAPT Recommendations for the Undergraduate Physics Laboratory Curriculum. Phys. Teach. **53**(4), 253 (2015). https://doi.org/10.1119/1.4914580
3. Day, J., Bonn, D.: Development of the concise data processing assessment. Phys. Rev. Spec. Top. Phys. Educ. Res. **7**(1), 010114 (2011). https://doi.org/10.1103/physrevstper.7.010114
4. Zwickl, B.M., Finkelstein, N., Lewandowski, H.J.: Development and validation of the Colorado learning attitudes about science survey for experimental physics. In: Engelhardt, P.V., Churukian, A.D., & Rebello, N.S. (eds.) 2012 Physics education research conference, vol. 1513, pp. 442–445. AIP, Philadelphia, PA, USA (2013). https://doi.org/10.1063/1.4789747
5. Wilcox, B. R., Lewandowski, H. J.: Open-ended versus guided laboratory activities: Impact on students' beliefs about experimental physics. Phys. Rev. Spec. Top. Phys. Educ. Res. **12**(2), 020132 (2016). https://doi.org/10.1103/physrevphyseducres.12.020132
6. Kontro, I., Heino, O., Hendolin, I., Galambosi, S.: Modernisation of the intermediate physics laboratory. Eur. J. Phys. **39**(2), 025702 (2018). https://doi.org/10.1088/1361-6404/aa9364
7. Hake, R.R.: Interactive-engagement versus traditional methods: a six-thousand-student survey of mechanics test data for introductory physics courses. Am. J. Phys. **66**(1), 64–74 (1998). https://doi.org/10.1119/1.18809

Evaluation of an Experimental Sequence on Introductory Quantum Physics Based on LEDs and the Photoelectric Effect

Massimiliano Malgieri⊙**, Pasquale Onorato**⊙ **and Anna De Ambrosis**⊙

Abstract In this chapter we describe an experimental sequence that helps students build a consistent mental model of the phenomena of photon absorption and emission. The sequence is based on a combination of experimental activities, such as the measurement of the Planck constant through the photoelectric effect and the threshold voltage of LEDs of different colors, and the measurement of the external quantum efficiency of LEDs. We discuss data from a final questionnaire for a sample of 147 18–19-year-old students. Results provide some encouraging indications in support of our approach, but also highlight limitations and allow us to uncover some non-trivial features related to students' knowledge and learning processes. The sequence of activities is suitable for visits and stages of high school students in University, but is also appropriate for the undergraduate laboratory.

Keywords Photoelectric effect · LEDs · Planck constant

Introduction

Experimental activities revolving around the Planck-Einstein relation for the photon, $E = h\nu$, based on either the photoelectric effect or light emission from LEDs, have been discussed several times in the literature [1–4]. The learning outcomes of such activities are somewhat less well studied. Although experimental activities are just one part of the educational process, a consolidated research tradition has argued the

M. Malgieri (✉) · A. De Ambrosis
Department of Physics, University of Pavia, 27100 Pavia, Italy
e-mail: massimiliano.malgieri01@universitadipavia.it

A. De Ambrosis
e-mail: anna.deambrosisvigna@unipv.it

P. Onorato
Department of Physics, University of Trento, 38123 Povo, Trento, Italy
e-mail: pasquale.onorato@unitn.it

© Springer Nature Switzerland AG 2019
E. McLoughlin and P. van Kampen (eds.), *Concepts, Strategies and Models to Enhance Physics Teaching and Learning*,
https://doi.org/10.1007/978-3-030-18137-6_10

importance of assessing the effectiveness of laboratory work in meeting its specific goals [5].

In the last few years the advancement of physics education research has provided important insights into what learning goals should be considered of primary importance in the initial teaching of photon emission and absorption phenomena. In Ref. [6] the author defines the main characteristics of conceptual change in students' interpretation of the photoelectric effect. Two main problematic issues are identified. The first is understanding that, according to the relation $E = h\nu$, the energy carried by the photon is proportional to its frequency (the frequency of the light used as source, if it can be considered monochromatic). The second is understanding that the intensity of light is proportional to the number of photons arriving on a given surface in a unit time, and, since the absorption of photons in electron energy transitions is a one-to-one mechanism with approximately fixed efficiency for a given material, the photocurrent intensity is proportional to the intensity of the incident light. The author develops a conceptual test comprising two questions to address the points described above; with some modifications, these two items correspond to the questions Q1 and Q2 of our final test, reported in the Appendix. Our test contains more items, which will be discussed in a future publication; in this chapter we limit ourselves to the discussion of the two questions derived from Ref. [6].

In 2016 a preliminary test of part of the sequence described in this chapter was performed with a single class of high school students [7]. The first version of the experimental sequence only included the measurement of the Planck constant through the photoelectric effect and the threshold voltage of LEDs. In the final test, most students (about 75%) answered correctly item Q1, demonstrating understanding of the relationship between the frequency of light and the maximum kinetic energy of the emitted electrons, including the presence of a threshold value. However, a majority of students (about 70%) answered "false" to item Q2. Most of the explanations for the wrong choice contained variations of the concept "because the photoelectric effect only depends on frequency".

In this article we describe a sequence of experimental activities aimed at providing students with an integrated understanding of the Planck-Einstein law governing photon emission and absorption, maintaining an equal focus both on the meaning of the energy-frequency relationship for a single photon and the relationship between light intensity and number of interactions per unit. The main strategy for achieving the latter objective is to guide students to explore not only the photoelectric effect, but also the light intensity versus voltage and light intensity versus current characteristics of LEDs, and estimate their external quantum efficiency. We discuss data from a final test on the $N = 147$ final year high school students who completed our sequence. We believe our data indicates a promising work direction, although "optimal" learning outcomes cannot be realistically achieved nor were expected in our setting, primarily because of the limited time available. The activities use sensors and software tools commonly available in undergraduate laboratories, and lasts approximately 5 h, thus it can be proposed to high school students within a full-day visit to the University labs. The activities are suitable both for final year high school and first year undergraduate students.

The Sequence of Activities

Measurement of Planck's Constant Through the Photoelectric Effect

The photoelectric effect was studied using the PASCO AP-9368 *h/e* apparatus, consisting essentially of a vacuum photodiode and a mercury gas vapor lamp to whose output slit a diffraction grating is applied, so that the different colors of the discrete spectrum of Hg can be separately directed towards the vacuum photodiode. The apparatus is different from the typical textbook setup for the photoelectric effect, in that the stopping potential is not provided by an external generator, but is gradually built through the charge unbalance between emitter and collector produced by the photoelectric effect. Thus, the apparatus does not rely on an externally applied voltage, which may reduce the possibility of interference with the belief that Ohm's law applies, or that the external voltage is responsible for producing the electron emission [8, 9]. Data collected are analyzed by each group and the relation

$$h\nu = eV_S + W_0 \tag{1}$$

between frequency ν, stopping potential V_S and work function W_0 is discussed. A schematic representation of the apparatus used by students is reported in Fig. 1.

Students also evaluate the threshold frequency and wavelength, and the work function of the material. The dependency between light intensity and number of emitted electrons is observed qualitatively by verifying that the diode potential takes more time to reach its equilibrium value for lower light intensities. The values of h obtained by students using the method based on the photoelectric effect differed 5–15% from the actual value (5.6–6.5×10^{-34} J s).

Measurement of the Planck Constant from the Turn-On Voltage of LEDs

The second part of the activity explores more in depth the meaning of the relationship $E = h\nu$ in the case of photon emission, with the interpretation of E as both the photon energy and the difference of energies involved in the electron transition (in the case of the LED, the bandgap energy). The basic properties of LEDs are explained to students through a simplified introduction to the theory of the p-n junction, in the form of a traditional chalk and blackboard lecture lasting 45 min–1 h. Although necessarily simplified, the treatment attempts not to omit critical details. Students then start working with their LEDs to obtain a new estimate of the Planck constant. Aiming at having them perform a consistent experimental procedure, we prefer not to

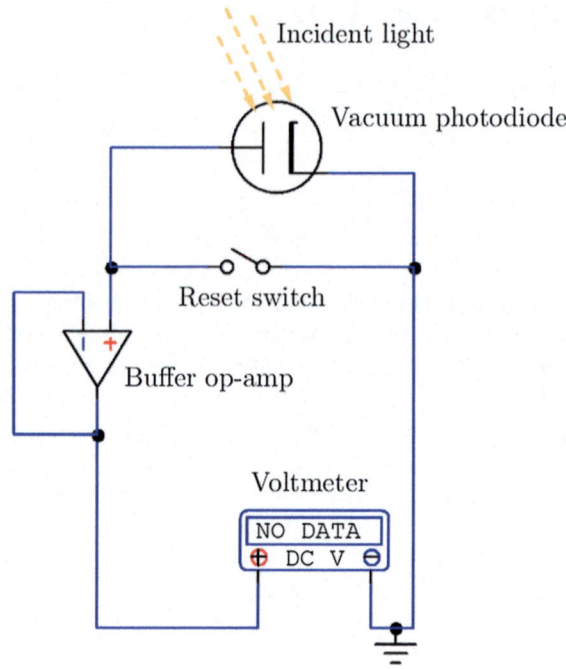

Fig. 1 Schematic representation of the PASCO apparatus; the op-amp in buffer configuration just serves to increase the isolation between the electrodes in presence of the voltmeter

let students force the intercept of their data fit to zero, as suggested by some authors. Rather, at the end of the discussion the resulting straight line is predicted as

$$V_0 = (h/e)v + \alpha(kT/e) \tag{2}$$

The constant α is left unspecified (except for the fact that it should be negative [10]) and is not measured; students concentrate on finding the slope of the straight line fit. The measurement is carried out in two steps.

- Under low light conditions, students first estimate the turn-on voltage of four LEDs (red, yellow/amber, green and blue) using their own eyes as light detectors, and the PASCO PS-2115 sensor: they measure and record the value of voltage when they see the first glimpse of light appearing in the LED capsule. The values of h found in this way typically range from 6.0 to 7.8×10^{-34} J s.
- Then students observe and record the current-voltage characteristic curve of the LEDs up to about 50 mA and provide a new estimate of V_0 by taking the intercept of a straight line fit of the last part of the data with the voltage axis. Students start with the same four LEDs used previously, but then add an infrared and an ultraviolet LED to their data set. The estimate of h using six LED is typically 5–15% larger than the true value (7.0–7.8×10^{-34} J s). However, students notice that five data points fit very well a straight line, while one (corresponding to the blue LED) is significantly displaced. By using only five LEDs students usually

find a value of h very close to the true one (6.5–6.9×10^{-34} J s). The discrepancy of the blue LED is discussed in terms of such devices being the product of a much newer technology, which required considerable research and which only partially fits with the simplified model presented.

Estimate of the External Quantum Efficiency

The focus of the third part of the activity is on connecting macroscopic quantities measured from a LED (the measured relationship between current and light intensities) to the microscopic emission processes (estimate of the external quantum efficiency). The main goal is to provide a quantitative demonstration of the relationship of direct proportionality between current and light intensity in the case of photon emission, which can then be extended, through the formation of a consistent mental model of electron-photon interactions, to what was observed qualitatively in the case of the photoelectric effect. The activity uses a predict-observe-explain approach, with students having to predict the shape of the illuminance-voltage curve based on their understanding of the LED model introduced. Students then measure such characteristic curve by using a PASCO-light sensor placed ad a fixed distance d over the capsule of a LED of their choice (among those emitting in the visible light range) together with the current-voltage sensor. They measure the curve up to a current intensity of about 25 mA finding that its shape is very similar to the current-voltage curve (Fig. 2, left).

Students are then asked to predict the shape of the illuminance-current curve, after which they visualize it in Capstone, finding that it is with good approximation linear (and a direct proportionality) in the considered range (Fig. 2, right). Illuminance

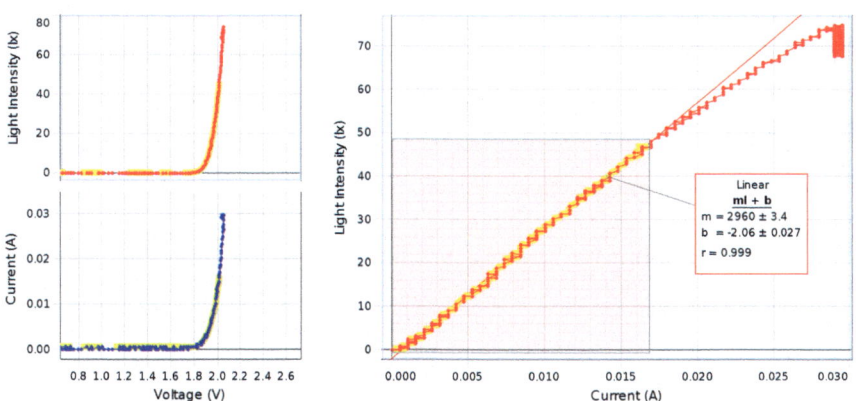

Fig. 2 Capstone screenshot of students' measurements. Light intensity versus voltage (red) and current versus voltage (blue) curves (left). Illuminance versus current curve, and its linear fit up to about 20 mA (right)

is then transformed into light intensity by using a table provided to students with an approximate conversion coefficient for each one of the LEDs. Laboratory sheets encourage students to articulate the idea that the direct proportionality between the measured light intensity and the current through the LED is due to the fact that a photon is emitted for each electron transition, and electron transitions happen with a certain, approximately fixed, probability. A parallel with the case of the photoelectric effect is explicitly made: in both cases, the interaction (absorption in the case of the photoelectric effect, emission in the case of LEDs) involves one photon and one electron with approximately fixed efficiency; thus, in both cases the result is a direct proportionality between current and light intensities.

Finally, students estimate the external quantum efficiency. At this aim, they multiply the angular coefficient of the fitted light intensity versus current intensity straight line by the factor $(\pi/4)d^2$, corresponding to considering uniform emission over a solid angle $\Omega = \pi/4$ (this is a rough approximation since the true emission pattern varies from LED to LED) thus finding the ratio of the total emitted power W_γ to the current i. This corresponds to finding

$$W_\gamma/i = n_\gamma h\nu/n_e e \tag{3}$$

where n_γ and n_e are the number of emitted photons and electrons respectively, per unit time. By inverting the above relation and inserting the value of ν corresponding to the LED chosen, students estimate the external quantum efficiency $\eta_{QE} = n_\gamma/n_e$. Although there is some degree of variability in students' results, in all cases they allow to place the LEDs in general categories such as "high efficiency" (ultra-bright/super-bright blue and red LEDs, measured $\eta \approx$ 20-40%), "medium efficiency" (yellow LED, measured $\eta \approx 1-3\%$) and "low efficiency" (green LED, efficiency of the order of $\eta \approx 0.1\%$ or less).

Methods

The sequence of activities was tested with 8 classes (N = 147 students) of the final year of the Italian high school course more specialized in science subjects ("Liceo scientifico"). In this type of school, the physics course in the second part of the year (roughly January to June) covers modern physics topics. The sequence was tested through a questionnaire, for which the first two items are reported in the appendix, and lasting about 45 min.

Results

Q1: Relationship Between the Frequency of Light and Maximum Kinetic Energy of the Emitted Electrons in the Photoelectric Effect

Overall, 91 students out of 147 (61.9%) provided the correct answer to the multiple choice item Q1. These results are reported in Table 1.

We also provide in Table 2 a phenomenographic categorization of the explanations offered by students who choose the correct items B. An explanation was judged as "correct and complete" if both the frequency threshold and the proportionality of the electron's kinetic energy to frequency were discussed. An example of such answer is "Until the frequency does not reach a minimum value, electrons are not emitted. Beyond that value, kinetic energy increases or decreases proportionally to frequency". Explanations of the correct choice discussing only the issue of proportionality or only the presence of a threshold are judged as basically correct, although not complete.

Q2: Relationship Between Intensity of Light and Number of Emitted Electrons per Unit Time in the Photoelectric Effect

Overall, 104 students out of 147, or 70.7% provided the correct answer "true" to question Q2. 39 students answered "false", and 4 students did not answer the question at all. Results are reported in Table 3.

The distribution of students' explanations for choices "true" and "false" is reported in Table 4. An example of a correct and complete answer is *"Higher intensity means more photons hit the pommel, thus a lesser time is needed for emitting a given number of electrons and the foils close quicker."*

Table 1 Results for item Q1

Option	Percentage of students
A	15.6
B (correct)	**61.9**
C	18.4
D	4.1

See the Appendix for the meaning of the options

Table 2 Students' explanations for answer B (N = 91) to item Q1, divided in categories

Type of answer	Number of students
Correct and complete (proportionality with threshold)	**61 (67%)**
Correct but not complete (proportionality only)	8 (8.8%)
Correct but not complete (threshold only)	4 (4.4%)
Incorrect or unclear	7 (7.7%)

Table 3 Results for item Q2

Option	Percentage of students
True (correct)	**70.7**
False	26.5
No answer	2.7

See the Appendix for the meaning of the options

Table 4 Students' explanations for answers "true" and "false" to item Q2, divided in categories

Type of answer	Number of students
Answer "true" (N_T = 104)	
Correct and complete (higher light intensity means more photons hit the zinc plate per unit time, more electrons are emitted from the plate per unit time)	**36 (35%)**
Correct but no mention of photons (higher light intensity means more electrons are emitted from the plate per unit time)	*32 (31%)*
Correct but no mention of electrons (higher intensity means more photons are carried by light/hit the plate per unit time)	*4 (4%)*
Only analogy with experiment in unclear terms	2 (2%)
Confuses light intensity with current intensity	5 (5%)
If intensity is higher golden leaves receive more energy	3 (3%)
If intensity is higher "stopping potential" is reached sooner	4 (4%)
Classical explanation (higher intensity wave gives more energy to each electron)	1 (1%)
Probable confusion with LEDs (emission of photons from electron transitions in the plate)	1 (1%)
Other unclear or undecipherable	3 (3%)
No explanation or restates the question	13 (13%)
Answer "false" (N_F = 39)	
Photoelectric effect does not depend on intensity but only on frequency	25 (64%)
The reverse is true (higher light intensity leads to lower speed)	1 (3%)
Light intensity only affects opening but not closing of leaves	1 (3%)
Other unclear or undecipherable	1 (3%)
No explanation or restates the question	11 (28%)

Overall, about half the students (72 out of 147, or 49%) provide the correct answer to Q2 while giving an essentially correct explanation (first three categories in Table 4).

Discussion

Our main result is that at the end of the experimental sequence both multiple choice items Q1 and Q2 were answered correctly by a majority of high school students (61.9% for Q1 and 70.7% for Q2). In both cases, about half the total number of students produced also a basically correct, although not always complete, explanation (50% for Q1 and 49% for Q2). Photoelectric effect is a difficult topic for students, and PER research has shown that instructors often underestimate the difficulty of the subject and overestimate their students' abilities [8]. In Ref. [6], items similar to Q1 and Q2 were proposed as a pre-test to first year physics course to $N = 129$ electronic engineering and astronomy undergraduates. However, Q1 only had two options (the correct one and a second one equivalent to option A or integration-like process) and for both questions the reason for the choice had to be chosen in a secondary multiple choice sub-item among five possible options. In this test the rate of correct answers (excluding the choice for the explanation) was 37.2% for the item similar to Q1 and 84.5% for Q2. If also the correct choice for the explanation is included, results read as 31.8% for the item similar to Q1 and 44.2% for Q2. A question similar to Q1 is also used in Ref. [8] as a clicker question in the context of a physics course for engineering majors; the authors consider the question as very difficult and depending on years they find results ranging from 36 to 56%. Given these previous results, and considering the composition of the sample (high school students) and the duration of the activity (5 h), we consider our results as valuable. Our data from a large sample of students and a renovated sequence of experiments, no longer show the anti-correlation effect between Q1 and Q2 which was found in a preliminary test [7]. In that context, learning that the kinetic energy of each emitted electron is independent of the light intensity seemed to inhibit the ability of a majority of students to answer Q2 correctly. However, our results can by no means be considered optimal and we still have indications that some degree of tension exists between the two issues. For example, as can be seen in Table 4, the percentage of students who justify their incorrect answer to item Q2 with variants of "photoelectric effect only depends on frequency" is still significant.

In the final activity on the quantum efficiency of LEDs, students are stimulated to provide an explanation of the proportionality between current and light intensities, which is not a trivial consequence of the formulas introduced, through a proper Predict-Observe-Explain approach. Besides some obstacles in the numerical calculations and conversions, students appear to understand the general meaning of what they are computing fairly well. In the following discussion, a connection with the case of the photoelectric effect is made, but this may not be sufficient for leading students to construct a unified mental model of emission and absorption phenomena. The sequence might benefit from being revised, perhaps adding a final activity which explicitly addresses the desired connection. For example, an experiment could be introduced in which one LED is used as emitter, and a second one is used as detector, exploiting the so-called "internal photoelectric effect".

Based both on spontaneous students' reactions (e.g. unsolicited final applauses at the end of the work by some of the classes involved, which in our experience are not common for this kind of activities) and teachers' feedback, the sequence had a very good reception and may have had some motivational impact on students who were considering studying physics at college.

Conclusions

We believe the sequence of experimental activities hereby presented to have the potential of improving students' understanding of the Einstein-Planck radiation law and of countering some significant difficulties. By performing experiments on both the photoelectric effect and photon emission by LEDs, a parallel can be drawn on salient characteristics of photon absorption and emission, which can be synthetically summarized as (a) in both cases the relationship holds for the energy carried by an individual photon and (b) in both cases the interaction is between an individual photon and an individual electron, with approximately fixed efficiency (probability) which produces a direct proportionality between current and light intensities. According to our data, the activity may help students gain a better understanding of the Planck-Einstein relationship, and improve their ability to distinguish the respective roles of the frequency and intensity of light. The activity was performed in the context of high school students' visits to the University labs; however, for the relevance of the concepts on which it focuses, the sequence is also suitable to be used in the undergraduate laboratory.

Appendix: Test Items

Item Q1

A photodiode like the one shown in Fig. 3, similar to the one used in the lab experience, is illuminated with light having a frequency slowly varying in time, from zero to a maximum value, as shown in Fig. 4.

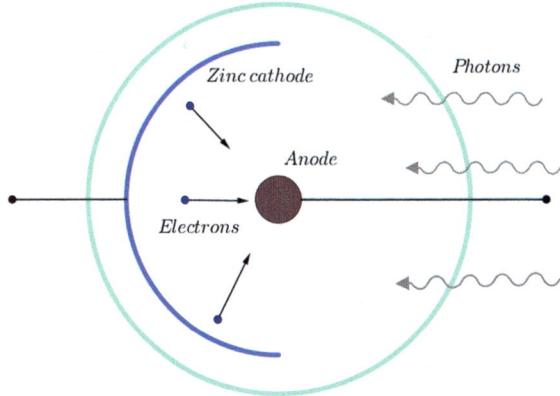

Fig. 3

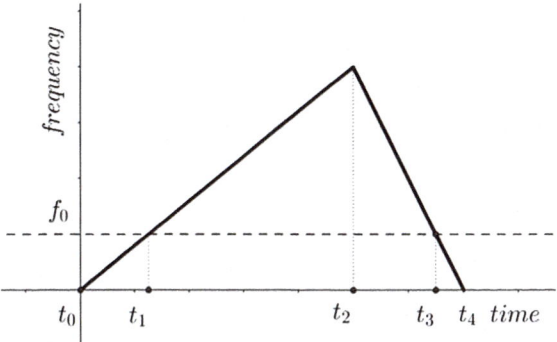

Fig. 4

Which one of the following graphs in your opinion best represents the graph of the maximum kinetic energy of the electrons emitted by the photocathode (in the graphs it is intended that when the kinetic energy is zero, no electrons are emitted)?

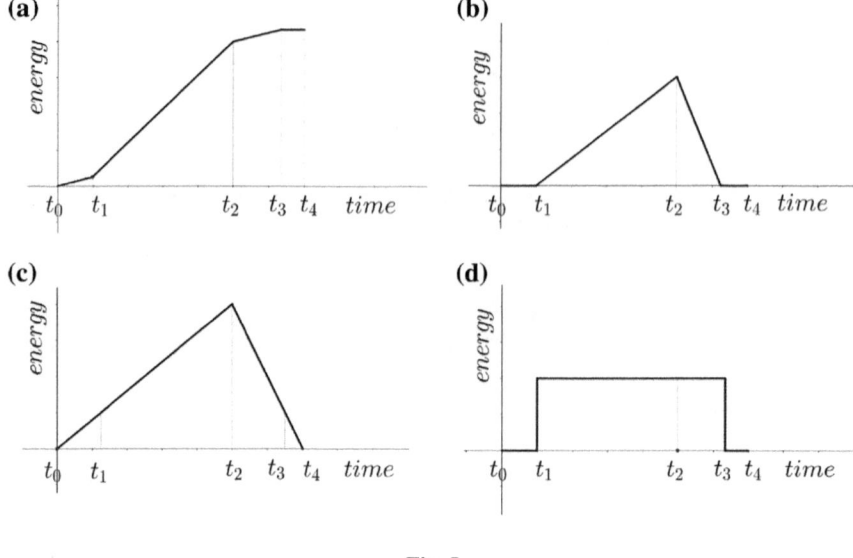

Fig. 5

Answer: _____
Explain your reasoning:

Item Q2

In the experiment represented in Fig. 6, a gold leaf electroscope is initially charged with a negative charge, so the gold leaves diverge. The upper pommel of the electroscope, which is usually used to charge or discharge the electroscope by contact or induction, is made of zinc, a metal which can produce the photoelectric effect if invested by ultraviolet light.

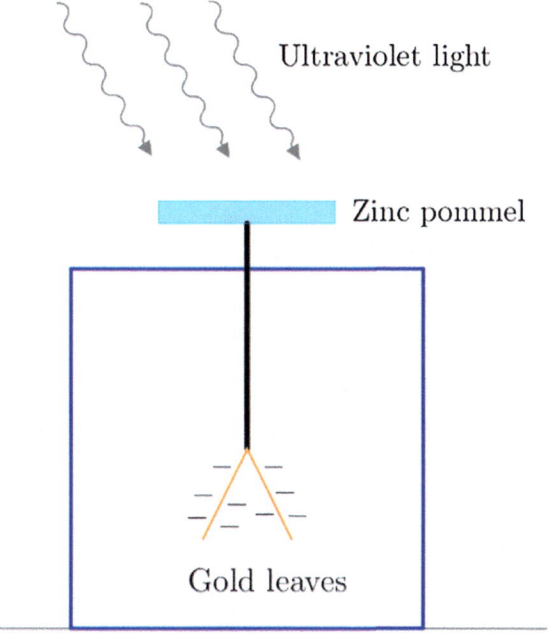

Fig. 6

If, like in Fig. 6, the zinc pommel is illuminated with ultraviolet light, zinc emits electrons and gradually the electroscope discharges, so that the two golden leaves return to their rest position (in which they no longer diverge). Assume that the UV lamp, which emits light at a fixed frequency, can be dimmed and used at different levels of intensity. Is it true that if the light intensity is higher the golden leaves close more rapidly, while if the intensity is lower, the leaves close more slowly?

True   False

Explain your reasoning:

References

1. Checchetti, A., Fantini, A.: Experimental determination of Planck's constant using light emitting diodes (LEDs) and photoelectric effect. World J. Chem. Educ. **3**(4), 87–92 (2015). https://doi.org/10.12691/wjce-3-4-2
2. Garver, W.P.: The photoelectric effect using LEDs as light sources. Phys. Teach. **44**(5), 272–275 (2006). https://doi.org/10.1119/1.2195395
3. Kraftmakher, Y.: Experiments with light-emitting diodes. Am. J. Phys. **79**(8), 825–830 (2011). https://doi.org/10.1119/1.3599072

4. Precker, J.W.: Simple experimental verification of the relation between the band-gap energy and the energy of photons emitted by LEDs. Eur. J. Phys. **28**(3), 493–500 (2007). https://doi.org/10.1088/0143-0807/28/3/010

5. Hofstein, A., Lunetta, V.N.: The laboratory in science education: Foundations for the twenty-first century. Sci. Educ. **88**(1), 28–54 (2004). https://doi.org/10.1002/sce.10106

6. Oh, J.Y.: Using an enhanced conflict map in the classroom (photoelectric effect) based on Lakatosian heuristic principle strategies. Int. J. Sci. Math. Educ. **9**(5), 1135–1166 (2011). https://doi.org/10.1080/02635143.2016.1222359

7. Malgieri, M., Onorato, P., De Ambrosis, A.: Assessing student's conceptual understanding in a laboratory on the measurement of the Planck constant. Talk at the GIREP 2016 seminar, August 30—September 3 2016, Krakow, Poland (2016)

8. McKagan, S.B., Handley, W., Perkins, K.K., Wieman, C.E.: A research-based curriculum for teaching the photoelectric effect. Am. J. Phys. **77**(1), 87–94 (2009). https://doi.org/10.1119/1.2978181

9. Steinberg, R.N., Oberem, G.E., McDermott, L.C.: Development of a computer-based tutorial on the photoelectric effect. Am. J. Phys. **64**(11), 1370–1379 (1996). https://doi.org/10.1119/1.18360

10. Pulfrey, D.L. Understanding Modern Transistors and Diodes. Cambridge University Press, Cambridge (2010). https://doi.org/10.1017/cbo9780511840685

Part III
Trends in Physics Teacher Education

The Role of Playing in the Representation of the Concept of Energy: A Lab Experience for Future Primary School Teachers

Alessandra Landini⑩, Enrico Giliberti⑩ and Federico Corni⑩

Abstract Energy, particularly in introductory physics at primary school level, is often taught in terms of list of different "forms of energy" and seldom as a unifying concept underlying many aspects of the world. However, the "substance" ontology for energy seems to be particularly productive in developing understanding of energy and energy transfers. From a methodological point of view, narratives and forms of "playing" are valuable and significant representations that allow learning scientific concepts. Through a physical experience, in the form of role play, we help developing the concept of energy flow/current and storage. In this contribution, we propose a laboratory activity in which future primary school teachers represent the process of energy exchange among energy carriers. The participants are required to study a simple toy, finding the energy carriers, and the role of each of them; additionally, they have to write a story, with as many characters as the energy carriers, telling how they exchange energy in the parts of the toy. Energy conservation and heat production are perceivable in the act of exchanging confetti which represent energy. The Energy Play helps the participants to visualize the energy as a substance, even though it is imperceptible. The analysis of the students' role plays and the information collected from questionnaires give feedback about students' conceptualization of some of the most significant aspects of energy.

Keywords Energy · Role-play · Quasi-material substance metaphor

A. Landini (✉) · E. Giliberti
University of Modena and Reggio Emilia, Viale A. Allegri, 9, 42121 Reggio Emilia, Italy
e-mail: alessandra.landini1970@unimore.it

E. Giliberti
e-mail: enrico.giliberti@unimore.it

F. Corni
Faculty of Education, Free University of Bozen-Bolzano, Via Ratisbona 16, 39042 Bressanone, Italy
e-mail: federico.corni@unibz.it

Introduction

In the field of science education, the complexity of the concept of energy has been highlighted in several studies centered on the difficulties of learning an "imponderable" idea. Moreover, the concept itself is very common inside and outside the scientific discourse [1]. As argued by Lancor: "The way energy is conceptualized varies depending on context [...] Scientists, generally, do not share a common language, even within a particular discipline [...]. The interdisciplinarity of energy issues in today's society means that special attention should be paid to differences in discourse between disciplines" [2]. Besides this complexity, the possibility of misconceptions arises [3, 4], stressing the need to look at a cross-disciplinary approach.

According to linguistic research, scientists were and are often able to investigate the multifaceted concept of energy, using models that employ conceptual metaphors [5, 6]. Furthermore, they usually formulate narratives of physical phenomena involving energy and explore these phenomena within the narrative [7–9]. In a recent article, Harrer notes that "the development of the modern energy concept is reviewed to show that the use of metaphors always has been and still is necessary for physicists to make sense of and communicate ideas about energy" [10]. Fundamentally, at the heart of a narrative about energy we can find different ontological metaphors and we can better uncover students' model of energy thanks to language: "Language does not refer directly to the world, but rather to mental models and components thereof! Words serve to activate, elaborate or modify mental models, as in comprehension of a narrative." [11]. This is the reason that moved us to take advantage of students' mental model and of the narrative way we all use to make sense of phenomena.

Theoretical Framework

We agree with Lancor when she says: "Viewing science as a set of coherent metaphors is not very different from thinking of science as a set of models; the way that we communicate scientific models is often metaphorical. Furthermore, multiple conceptual metaphors may be necessary to describe one scientific model, as is the case with energy." [2]. For this reason, we decided to focus on conceptual metaphor studies [12–14] and embodied cognition. The cognitive and learning value of using metaphor and analogy to understand a target concept are well-established: according to Treagust and Duit [15] metaphors and analogies "permeate all discourses, are fundamental to human thought and are not simply teaching tools", they are "a fundamental principle of thought and action" [16].

Lancor [2] studied students' ideas on energy by analysing examples of metaphorical language and explicit analogies, according to the theory of the embodied cognition of Lakoff and Johnson [12, 13]. The goal was to underline the use of common conceptual metaphors across different disciplines. She describes different energy mapping as follows:

- energy as a substance that can be accounted for
- energy as a substance that can change form
- energy as a substance that can flow
- energy as a substance that can be carried
- energy as a substance that can be lost from a system
- energy as a substance that can be stored, added or produced

The first mapping, describing energy as a substance residing in "containers", is very common in various disciplines: it is reinforced through graphical representation and, as the author summarizes, gives scientists "a tool to apply energy conservation quantitatively", tracking energy changes at the same time. The metaphor of "ENERGY AS A SUBSTANCE" facilitates the concepts of conservation and transfer aspects of energy.

The third metaphor, "energy flows", is often used in textbooks. Lancor highlights the usefulness of this metaphor, talking about energy transfer in a system: "Thus this metaphor highlights the transfer of energy downplaying energy transformation". She concludes valuating this metaphor as a convenient way to discuss continuous, uniform energy transfer through a system.

The fourth metaphor focalizes on the substance contained and carried through energy carriers. Lancor argues: "it is more scientifically accurate to view an energy transformation as energy being transferred from one carrier to another. (...) The energy has a different carrier". This is right also for Falk et al. when they suggest that it is more scientifically accurate rather than thinking of energy as changing form [17].

Lancor's studies emphasize the models and the metaphoric framework of students' comprehension of energy, taking into account that they form "a set of coherent conceptual metaphors for energy" [2]. Although the limitations of the substance metaphor are well recognized in several studies [18–20], Lancor supports the idea to use this metaphoric framework as a formative assessment tool, useful to monitor students' conceptual development.

This formative assessment should be a part of the teaching-learning process. Metaphors are sometimes difficult to interpret, but they are very useful in the evaluation process: the students' ideas of energy and the underlying substance metaphor could be considered as a way of evaluating their ideas on this complex concept. For this reason, it is essential that teachers are aware and make use of these aspects in the evaluation process of pupils.

Close and Scherr [21] selected the substance metaphor for energy "as a primary focus of our instruction because of its advantages teaching conservation, transfer and flow". According to the author it shows the following features:

- energy is conserved;
- energy is localized;
- energy is located in objects;
- energy can change form;
- energy is transferred among objects and energy can accumulate in objects.

The authors claim: "These features constitute a powerful conceptual model of energy that may be used to explain and predict energy phenomena".

Purpose and Methods of the Study

While agreeing with Close and Scherr [21] on the ontological choice, in this research activity we highlight some considerations about instruments that could be used to allow a slightly different representation of energy during the teaching-learning process.

To facilitate this process, teachers need a good experiential project. In our proposal, this project should be:

- coherent with the above set of metaphors;
- based on the idea of embodied cognition.

We are suggesting here a set of training activities, addressed to pre-school teachers, in the form of Research Based Learning, in which a narrative approach to energy, together with a well-built context of playing and embodied simulation, allows teachers to have experiences that can lead to a good formalization of the concept of energy. The aim is a conceptualization of energy that explains everyday experience and, at the same time, fosters a deep scientific comprehension of energy.

We have already presented a revised use of Energy Flow Diagrams, assessed during an innovative path for the energy concept comprehension [22–25]. This cognitive and didactic tool, based on the quasi-material substance metaphor, and figuratively representing the metaphorical aspects of energy through the natural language, allows a narrative process of explanation supported by graphic symbols. The symbols, specifically designed to represent the energy flow through energy carriers, seem to support a better use of the language according to energy features as a fluid substance [25].

We suggest the image of a substance-like conserved quantity transfer to account for the proportionality between the quantity and intensity drop of the "acting" force (available energy) and the quantity and intensity rise of the "driven" force (absorbed energy). This use of the language and the related natural metaphorical expressions seem to incorporate the aspects of conservation, accumulation, transportability and transferability of energy. Furthermore, energy is introduced because of the need of accounting for phenomena where two natural phenomena interact in a device (e.g. wind and rotation in a windmill as in Fig. 1), whose operation can be narratively and formally described.

The purpose of this study is to analyse a specific use of these energy flow diagrams and a form of role-play (as a narrative simulation of processes), connected to the fluid substance metaphor and the theory of embodied cognition, both to suggest students' comprehension and to investigate their conceptualization of energy. In this sense, the role-play is seen as a metacognitive context for the expression of the mind through

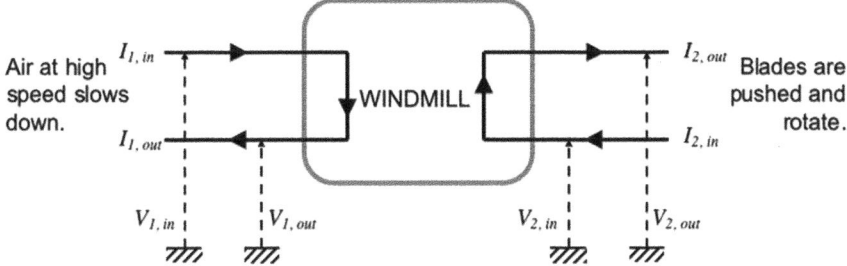

Fig. 1 Air and angular momentum interact in a windmill

the body, in which the language becomes fertile ground for the authentic evaluation of the understanding of natural phenomena.

The specific research question addressed in this study is: «To what extent does the metaphorical use of language and the related role-play affect students' comprehension of energy features?»

The science literature and linguistic and cognitive studies outlined above led us to hypothesize that the development of understanding of the energy concept could be facilitated by the quasi-material substance metaphor and confident use of language during the laboratories [26]. Furthermore, we hypothesized that specific use of diagrams connected to narratives and a specific use of role-play, connected to sensory-motor experience, could be conceptually productive.

In terms of Formative Assessment for Learning [27] we decided to analyse role-plays realized by pre-service teacher students of the Faculty of Education in Bressanone (UNIBZ) during an energy lab, coherent with the theoretical framework presented above. The study consisted of a 4 h activity for 59 students in their first year of university who had not completed their planned physics course yet.[1] The purpose of the lab for the students was to produce role-plays concerning energy for didactic use: after a narrative introduction to the topic through the vision of an animated tale, students studied a toy with its parts and mechanisms in groups and were asked to identify energy carriers and their roles. Then they wrote a narration, with as many characters as the energy carriers, telling how they act and exchange energy in the parts of the toy. Finally, students were asked to represent the energy carriers in a play, with confetti playing the role of the energy carried. The focus of the representation was energy and its transfer among carriers through different exchanges, showing the effects on qualitative aspects of the carriers. As a final activity, students viewed and commented all the role-plays.

Students' role-plays were then analysed. The research included two main steps:

1. role-play analysis, from the narrative and metaphorical point of view;

---

[1] The duration of the laboratory was dictated by organizational needs of the university to which the research group adapted this very first experimentation. Following the experimentation, which gave encouraging results, the laboratory was proposed in a training course for pre-service teachers in the third year of their Masters degree with a total duration of 16 h.

2. analysis of students' conceptualizations using video, integrating language analysis with gesture analysis.

In addition, evidence of the development of students' understanding of energy was collected in two different ways: by a questionnaire set after the Lab experience and by logbooks (cognitive autobiography) set one week later. The two assessment tools will be the subject of further study.

In this work we will specifically present the results of the role play and of a questionnaire analysis to emphasize the use of a Narrative Assessment for Learning to understand students' conceptualizations and to investigate their metaphorical use of language and gestures.

Role-Play as Incarnate Simulation: A Possibility to Make Energy "Ponderable"

The role-play generally involves a situation, that is a dynamic context where at least two subjects can be distinguished, with their role relationships, into a place, concerning a topic. In our case the agents are the energy carriers and the energy exchangers who, with their role relationships, into a system/space and their reactions, share the situation where the energy (confetti) is transferred.

For each situation/story it is necessary to identify [28]:

- where (place/part of the system) it happens,
- when it happens (time),
- who happens to (carriers/agents; exchangers/patients),
- what relationships are there (correlations/relationships with potential differences),
- what it is (situation/problem).

The advantages of our Energy Play are on different levels:

- on the semiotic level, this narrative technique combines words with facial expressions, gestures, and objects, which are actually manipulated or suggested by the gesture itself, take shape and life;
- on the psychological level, being a simulation, it is experienced as a game, stimulating an active and emotional participation;
- on the neurocognitive level the incarnate simulation and the "physicalization" of inanimate objects, activates sense-motor areas of the brain that asymmetrically refer to other areas of language and understanding;
- the social level also falls into role-play comprehension mechanisms: the latter can be a kind of apprenticeship of relational and emotional models and serve as experiential context for understanding reality.

Analysis and Results

This work on student role-play concerning energy arose out of studies regarding the relationship of mind with body and studies on representation. Moreover, this research is designed to examine a much broader spectrum of energy features, including the contribution of gesture, bodily perceptions and interaction. The story and the number of sequences are necessarily small, but the theoretical structure used in this analysis can serve as an example of integrated reading, where oral and body narration make a contribution to describing the procedures and the conceptualization processes in progress. We have distinguished some nodes of interest to evaluate the conceptualization of energy according to the ontology of the fluid substance: Were all parts of the energy diagram there? Which carriers were present? Which energy exchangers were present? Was there a relation that showed the difference of potential before and during the exchange? Does energy conservation emerge? Is the metaphor of the fluid substance present?

The following pages show a role-play analysis for a "Windmill generator"[2] and some considerations about the results of this "narrative incarnate simulation". Table 1 is the result of a qualitative analysis autonomously realized by three researchers and then shared according to the criteria outlined above.

The results obtained, relating to the research questions, can be summarized as follows:

- all the carriers are present and correct;
- all exchangers are present, but during the role-play, we found that they were acted more as agents rather than simply mechanisms that allow the energy exchange;
- the difference in potential is not significant.

The use of the energy flow diagram, connected to the metaphorical use of language and the related role-play, seems to affect students' comprehension of the energy features, but we could not find a clear conceptualization of potential difference. The idea of energy conservation is present.

In this paper, we cannot provide a full analysis: semiotic, psychological and social factors would deserve a more thorough and deeper discussion. However, two themes that are at the basis of the theoretical framework need to be emphasized: the connection between natural language use and the use of the body in the narrated simulation. From the linguistic point of view, students' concepts of energy appear to arise from narration, especially in the form of quasi-material substance metaphor; moreover, the analysis draws attention to a multi-metaphorical space around the fluid substance metaphor: More is up, Changes Are Movements, Causation, and Part/Whole schemata, to name but a few. Integrating oral and visual metaphors and gestures

---

[2]The energy flow in the toy can be outlined as follows: First Carrier: Free Air/Wind—Exchanger: Rotor—Second Carrier: Angular Momentum—Exchanger: Gears—Third Carrier: Angular Momentum—Exchanger: Dynamo—Fourth Carrier: Electricity—Exchanger: Lamp—Fifth Carrier: Light.

Table 1 Video sequence from 1 to 7

Sequence	Narration and linguistic indicators	Gestural indicators	Energy conceptualization as a quasi-material substance
1st sequence	*The wind is full of force and will to play* The potential is high (the wind is a container of a fluid substance; it is associated with a spatial location, even if it seems spread out)	Two students are smiling, keeping confetti in the hands, dancing excited (the emotion is in the container and the level is up-more is up)	The potential is high. (energy as an amount of substance in a container) The carrier MOTION (wind) seems to have a certain amount of energy in itself
2nd sequence	*It (wind) meets a nice sleeping rotor and he desired to play with him* (change of level: from sleeping to awake) (causation: playing with the rotor causes the rotor to wake up)	The two students look at the rotor, they look each other, smile, and start to move forward to the rotor (causation: the wind provides the effect on the rotor, which it will turn)	The carrier MOTION (wind) moves to the rotor (energy flows) The wind brings energy to the rotor (energy is transferred)
3rd sequence	*The Wind runs to meet him [the rotor] and the blades, crazy of joy, started to rotate* (causation: the blades turns because of the wind) (change of potential: from low speed to high speed)	Students head off quickly against the blades, represented by four other students, who, happy, begin to spin on themselves, simulating the rotation of the rotor and bringing the confetti. The wind goes away. As the four students turn around, they drop a little amount of confetti (causation) (exchange of confetti) (loss of confetti) (the carrier WIND disappears)	MOTION brings energy from LINEAR MOMENTUM to the ANGULAR MOMENTUM (Flow of energy from the wind to the blades; the energy flows through the exchanger (blades) the energy flows There is a loss of energy during the exchange)

(continued)

Table 1 (continued)

Sequence	Narration and linguistic indicators	Gestural indicators	Energy conceptualization as a quasi-material substance
4th sequence	*She did not notice, however, that there were two very hungry gears next to her, so hungry to devour much of the energy* (interaction) (energetic state as a resource)	Three students begin to rotate representing the largest toothed wheel, two other students simultaneously move and represent the smallest wheel: while turning, they pretend to eat the confetti and lose a bit of it (interaction) (confetti are transferred from one gear to the other) (energetic state is a possession)	Flow of energy through the gears: The energy flows through the exchanger-gears. During the transfer there is a loss of energy
5th sequence	*Full and tired of eating, they gave part of the energy to their dynamo friend, who slowly began to move her fingers, producing electricity* Causation (whole/part metaphor)	The gears, represented by the five girls, simulating satiety and fatigue, pass the confetti to the dynamo represented by three crouched girls, who stand up thanks to the confetti (interaction) (confetti are transferred from one gear to the dynamo)	There is a decrease in speed through the gears, due to friction The energy flows thanks to the ANGULAR MOMENTUM through the dynamo. The angular momentum has a high potential, but we cannot notice the difference in potential after the exchange

(continued)

Table 1 (continued)

Sequence	Narration and linguistic indicators	Gestural indicators	Energy conceptualization as a quasi-material substance
6th sequence	*The generous and gentle electricity decides to accompany her friend energy to the doxy bulb* (desire is moving: the electricity flows to the doxy bulb)	Two students, representing electricity, receive confetti from the dynamo; some confetti are lost The electricity seems to have high potential (students are again excited) quickly bring it to an immobile student, representing the light bulb Interaction Confetti are transferred from the dynamo to the bulb (potential change from low to high)	The ELECTRICITY carrier receives the energy from the ANGULAR MOMENTUM carrier through the dynamo Energy is carried by electricity Some energy is lost during the transfer
7th sequence	*The bulb happily animated, generating light and illuminating the whole world* (causation) (Change of potential from low to high)	Through the lamp you can see confetti flowing to the last student, who impersonates the final carrier, the light. The confetti, arrived at the end of the path, are less than those at the beginning of the journey, because a part of them has been lost in the transfer from one carrier to the other. However, the total amount of confetti present on the scene remains the same (interaction) (source-path-goal schema) (part/whole schema) (real value is substance) (balance schema)	Through the lamp the electricity carrier makes the energy flow to the LIGHT While flowing through the lamp, the energy decreases again, which causes the production of entropy The amount of energy that enters into the system is the same amount of energy that comes out of the system Electricity brings energy to light through the bulb Some energy is lost in the transfer Conservation of the amount of energy

seems to enrich the experience: different Image Schemas are connected, and this seems to create inter-textual and intra-textual coherence in the narration [29].[3]

In the first sequence, for instance, the narrator says: "The wind is full of force and will to play"; the underlying analysis is "The wind is a container of a fluid substance; it is associated with a spatial location, even if it seems to be spread out". At the same time narration implicitly focuses on "Energy as an amount in a container". This quantitative aspect has been integrated by gesture and facial expressions, which stress the qualitative aspect of the phenomena: Two students are smiling, keeping confetti in their hands, dancing excitedly (the emotion is in the container and the level is up-More is up). This seems to support the idea that the carrier, air in motion, has a certain amount of energy in itself with a high level of potential.

In order to evaluate the functionality of the laboratory with respect to the development of the scientific concept of energy, self-assessment questionnaires were given to the 59 students attending the laboratory. We will show the analysis of the answers related to the question: "Describe briefly what happened to the energy-confetti during the playing that you have represented together with your companions". We analysed the answers/narratives in qualitative terms, based on three main themes: (1) Presence of the Fluid substance paradigm; (2) Presence of carriers and exchangers; (3) Conservation of energy. In summary, the data show:

- theme 1 present in 26 narratives out of 59
- theme 2 present in 36 narratives out of 59
- theme 3 present in 32 narratives out of 59

Seventeen narratives comprise all three themes, and 26 contain at least two of the themes, bringing to more than half the number of narratives with elements of the chosen energy paradigm. We would like to highlight how in a 4 h workshop it was possible to involve first-year students who had not attended the Physics course yet in the conceptualization of such a complex subject. The paradigm presented, although difficult, interested them and has been productive for the initial conceptual development, which has brought new and complex themes in half of their narratives, while in the 29% of student's stories all three main themes have been introduced.

---

[3]Kovecses says: "I suggest that image schemas, domains, and frames are all conceptual structures that can be found at what, in previous work, I called the supraindividual level. This is the realm that informs the basic ontology of conceptual systems. These are the structures that we have in long-term memory and that provide the conceptual substrate of meaning in general and meaning in language in particular".

When we embed these structures in an embodied narration, in a "communicative" representation of a phenomenon, we think to perceive another level of structure; Kövecses adds "we put this huge amount of tacit knowledge to use in order to achieve particular goals (social, expressive, rhetorical, etc.). The job is performed online by individual speakers in specific contexts who manipulate and modify the conceptual structures in long-term memory according to their communicative goals. The conceptualization process and the language that is used are, in this case, fully contextualized. It is at this level that we utilize mental spaces, or, equivalently, scenes or scenarios, as suggested by Musolff, that are not part of our routinely used cognitive and linguistic repertoire".

Conclusions

Energy role-plays are useful to enhance the idea of energy as a substance and, at the same time, to evaluate the students' level of conceptualization: moreover, they help to visualize the flow of energy together with the energy carriers and to "show" the interaction between the carriers in terms of change of potential. These findings support the idea that embodied simulations and "physicalization" of inanimate objects can improve students' comprehension of energy features. Energy Play connect the simulation to the metaphorical conceptualization and affectively engage the actors, creating a contextualised and communicative situation.

Understanding energy as a substance, unaltered and distinct from both the carriers and the exchangers, could avoid confusing conceptualization of this imponderable and important subject. In such a way, we suggest that the concepts of *energy flow* and *conservation* should be introduced after a specific training activity to primary teachers and then to children, followed by other aspects of energy, deepen through incarnate simulation and dynamic representations in the form of Energy Play.

References

1. Lemke, J. L.: Talking Science: Language, Learning, and Values. Ablex Publishing, Stamford (1990) (See also Lemke, 1997, Spanish edition)
2. Lancor, R.: Using metaphor theory to examine conceptions of energy in biology, chemistry, and physics. Sci. Educ. **23**(6), 1245–1267 (2014). https://doi.org/10.1007/s11191-012-9535-8
3. Carey, S.: The Origin of Concepts. Oxford University Press (2009)
4. Wiser, M., Amin, T.: Is Heat Hot?" Inducing conceptual change by integrating everyday and scientific perspectives on thermal phenomena. Learn. Instr. **11**(4–5), 331–355 (2001). https://doi.org/10.1016/S0959-4752(00)00036-0
5. Duit, R.: On the role of analogies and metaphors in learning science. Sci. Educ. **75**(6), 649–672 (1991). https://doi.org/10.1002/sce.3730750606
6. Collins, A., Gentner, D.: How people construct mental models. In Holland, D., Quinn, N. (eds.) Cultural Models in Thought and Language. Cambridge University Press, Cambridge, pp. 243–265 (1987). https://doi.org/10.1126/science.240.4855.1080
7. Fuchs, H. U.: From image schemas to dynamical models in fluids, electricity, heat, and motion. Phys. Educ. Res. (2007). Retrieved from https://home.zhaw.ch/~fuh/LITERATURE/Literature.html
8. Fuchs, H. U.: Force dynamic gestalt, metaphor, and scientific thought. In: Proceedings of "Innovazione nella didattica delle scienze nella scuola primaria: al crocevia fra discipline scientifiche e umanistiche", Ed. Artestampa, Modena (2011)
9. Kubli, F.: Teaching as a dialogue—Bakhtin, Vygotsky and some applications in the classroom. Sci. Educ. **14**(6), 501–534 (2005). https://doi.org/10.1007/s11191-004-8046-7
10. Harrer, B. W.: On the origin of energy: metaphors and manifestations as resources for conceptualizing and measuring the invisible, imponderable. Am. J. Phys. **85**(6), 454–460 (2017). https://doi.org/10.1119/1.4979538
11. Hestenes, D.: Notes for a modeling theory of science cognition and instruction. In: Proceedings of the 2006 GIREP conference: Modelling in Physics and Physics Education. University of Amsterdam, Amsterdam, Netherlands (2006). Available at: https://pdfs.semanticscholar.org/3a96/f94fd0da55777df19980593ef17d87397878.pdf

12. Lakoff, G., Johnson, M.: Metaphors We Live By, 2nd edn. University of Chicago Press, Chicago (1980)
13. Lakoff, G., Johnson, M.: Philosophy in the Flesh. Basic Books, New York, NY (1999)
14. Lakoff, G.: Mapping the brain's metaphor circuitry: Metaphorical thought in everyday reason. Front. Hum. Neurosci. **8**, 958 (2014). https://doi.org/10.3389/fnhum.2014.00958
15. Treagust, D.F., Duit, R.: On the significance of conceptual metaphors in teaching and learning science: commentary on Lancor; Niebert and Gropengiesser; and Fuchs. Int. J. Sci. Educ. **37**(5–6), 958–965 (2015). https://doi.org/10.1080/09500693.2015.1025312
16. Niebert, K., Marsch, S., Treagust, D.F.: Understanding needs embodiment: a theory-guided reanalysis of the role of metaphors and analogies in understanding science. Sci. Educ. **96**(5), 849–877 (2012). https://doi.org/10.1002/sce.21026
17. Falk, G., Herrmann, F., Bruno, Schmid G.: Energy forms or energy carriers? Am. J. Phys. **51**(12), 1074–1077 (1983). https://doi.org/10.1119/1.13340
18. Amin, T.: Conceptual metaphor meets conceptual change. Hum. Dev. **52**, 165–197 (2009). https://doi.org/10.1159/000213891
19. Scherr, R., Close H.G., McKagan, S.B., Vokos, S.: Representing energy. I. Representing a substance ontology for energy. Phys. Rev. Spec. Top. Phys. Educ. Res. **8**, 020114 (2012). https://doi.org/10.1103/physrevstper.8.020114
20. Scherr, R., Close H.G., Close, E.W., Vokos, S.: Representing energy. II. Energy tracking representations. Phys. Rev. Spec. Top. Phys. Edu. Res. **8**, 020115 (2012). https://doi.org/10.1103/physrevstper.8.020115
21. Close, H.G., Scherr, R.E.: Enacting conceptual metaphor through blending: learning activities embodying the substance metaphor for energy. Int. J. Sci. Educ. **37**(5–6), 839–866 (2015). https://doi.org/10.1080/09500693.2015.1025307
22. Fuchs, H.U., Corni, F., Giliberti, E., Mariani, C.: Force dynamic gestalt of natural phenomena: teaching the concept of energy. In: E-Book Proceedings of the ESERA 2011 Conference: Science learning and Citizenship, pp. 31–37. ESERA, Lyon (2012)
23. Corni, F.: An approach to the concept of energy for primary school: disciplinary framework, elements of a didactic path and assessment scale. In: SEENET-MTP Seminar for Teachers: Trends in Modern Physics (2011)
24. Corni, F., Giliberti, E., Mariani, C.: The MLE-energy software for energy chains modelling. In: MPTL 14 Proceedings, Udine, 23–25 Sept 2009
25. Altiero, T., Bortolotti, C.A., Corni, F., Giliberti, E., Greco R., Marchetti, M., Mariani, C.: Introduzione elementare all'energia: un laboratorio di scienze per insegnanti di scuola primaria. In: Menabue, L., Santoro, G. (eds) New Trends in Science and Technology Education. Selected Papers, Modena, Italy, 21–23 Apr 2009, pp. 157–170, Clueb, Bologna (2010)
26. Landini, A., Corni, F.: Dalla narrazione all'esperienza in laboratorio: giochiamo e ragioniamo sull'Energia. L'educazione permanente a partire dalle prime età della vita. Franco Angeli Milano, Italy. pp. 1059– 1070. Conference Proceedings (2016)
27. William, D.: Embedded Formative Assessment. Solution Tree Press, Bloomington, USA (2011)
28. Zoletto, D.: Dai giochi del far finta ai giochi di ruolo e di simulazione. www.fisica.uniud.it/URDF/masterDidSciUD/materiali/pdf/zoletto03.pdf, Università degli Studi di Udine (2003). See also Bondioli, A.: Gioco e educazione. Franco Angeli, Milano (1996)
29. Kövecses, Z.: Levels of metaphor. Cogn. Linguist. **28**(2), 321–347 (2017). https://doi.org/10.1515/cog-2016-0052

Development and Assessment of Inquiry Skills in the Context of SSI with Pre-service Teachers

Ruth Chadwick⬤, Eilish McLoughlin⬤ and Odilla E. Finlayson⬤

Abstract Recent research has focused increasingly on student-led inquiry into socio-scientific issues (SSI) as a way of developing the skills and knowledge of science in secondary school students. However, there are barriers for teachers implementing this type of inquiry such as lack of pedagogical skills. This study explores the experiences of pre-service teachers (PSTs) as they engage in student-led, research-based inquiry in the context of SSI relating to the societal implications of a technological application of science. It is proposed that through engagement in this type of inquiry as learners that the PSTs will develop the skills and knowledge of science and overcome some barriers to implementing inquiry in their own teaching practice. 43 PSTs completed an open-ended questionnaire relating to their experience and their answers were subject to thematic analysis. Findings indicate that the PST experience focused on development of research and presentation of evidence. When demonstrating their knowledge, the PSTs did so within the context of the SSI. Implications for the implementation of inquiry in the context of SSI in secondary schools and for initial teacher education are discussed.

Keywords Socio-scientific issues · Inquiry · Pre-service teachers

R. Chadwick · E. McLoughlin (✉)
School of Physical Sciences, Centre for the Advancement of STEM Teaching and Learning, Dublin City University, Dublin, Ireland
e-mail: Eilish.McLoughlin@dcu.ie

R. Chadwick
e-mail: ruth.chadwick3@mail.dcu.ie

O. E. Finlayson
School of Chemical Sciences, Centre for the Advancement of STEM Teaching and Learning, Dublin City University, Dublin, Ireland
e-mail: Odilla.Finlayson@dcu.ie

© Springer Nature Switzerland AG 2019
E. McLoughlin and P. van Kampen (eds.), *Concepts, Strategies and Models to Enhance Physics Teaching and Learning*,
https://doi.org/10.1007/978-3-030-18137-6_12

Introduction and Research Questions

Recent research has focused increasingly on student-led inquiry into socio-scientific issues (SSI) as a way of developing the skills and knowledge of science in secondary school students. SSI are scientific topics with societal implications that can be used as the contexts for learning [1]. However, there are barriers that teachers face when facilitating inquiry in the context of SSI. Teachers are said to be unclear about the pedagogy and assessment in relation to inquiry teaching and that this is a particular issue for pre-service teachers (PSTs) [2, 3]. PSTs may have personally been taught science in a didactic, rather than inquiry approach and therefore may be conditioned, through their own educational experience, to use a more didactic approach with a focus on gaining content knowledge of science [4]. In order to increase teacher confidence in teaching through inquiry, and counteract the possible conditioning towards didactic approaches, PSTs should be involved in learning activities, through initial teacher education, relating to inquiry in the classroom [2].

This study explores the PSTs' experiences of the module and asks: What are the PSTs' experiences of carrying out inquiry in the context of SSI as learners?

The student-led inquiry was carried out as part of the initial teacher education program for second year undergraduate PSTs of math and science and aimed to develop skills and knowledge of science in PSTs. It is proposed that through experience of inquiry methods as learners, the PSTs will not only develop the skills and knowledge of science but also begin to reconceptualize some of the perceived barriers discussed above.

The inquiry was based on the Irish Junior Cycle Science in Society Investigation [5] and was carried out by the PSTs over 4 h of contact time, consisting of two phases:

1. Phase one: Research
2. Phase two: Communicating findings

PSTs carried out research for three hours, in class, with access to the internet. During this stage, PSTs chose their topic, decided their specific research question and gathered and recorded their research information. Collaboration was permitted during this phase. This was carried out during lab time. The PSTs were then given one week to produce an electronic document containing their research information and sources. The SSI context was given to the PSTs and related to the scientific, societal and environmental implications of a technological application of physics [6]. However, an element of choice was introduced as the PSTs chose their own question for research. The PSTs were then given 1 h to communicate their findings, using the information gathered during the research phase. PSTs were permitted to communicate their findings in a format of their choice. The majority of the PSTs chose either a written report or a presentation which they prepared during lab time.

Methodology

The study was conducted over two years with two cohorts of PSTs, 43 students in total (23 in year one, 20 in year two). The PSTs were in the second year of a concurrent B.Sc. in Science Education that qualifies them to teach mathematics, chemistry or physics at secondary level. The researcher facilitated the module and assessed students. Qualitative data relating to the PSTs' experience of carrying out inquiry in the context of SSI as learners was collected through an open-ended questionnaire (Table 1) and thematic analysis was carried out on the PSTs' responses [7].

Findings

Table 2 shows the results of thematic analysis of questionnaire data relating to the PST experience of inquiry in the context of SSI. The sub-themes are displayed in order of the number of references to the sub-theme from highest to lowest.

Table 1 Questionnaire relating to PSTs' experience of carrying out carrying out inquiry in the context of SSI as learners

1. What do you think are the learning intentions and success criteria of the Science in Society Investigation in Science?
2. From your experience of carrying out the Science in Society Investigation, list the top 3 things you learnt? This may be knowledge or skills or something else. Give an example for each.
3. What about the Science in Society Investigation went particularly well for you? Give examples.
4. What about the Science in Society Investigation was particularly challenging for you? Give examples.
5. If you had the chance to complete the Investigation again, what changes, if any, would you make to how you carried out your Investigation?

Table 2 PST experience of inquiry in the context of SSI

Theme	Sub-theme
Skills	Research
	Present information
	Propose investigatable questions
	Self-management
Knowledge	Implications of scientific knowledge for society
	Recall and apply content knowledge

The PSTs' experience focused on the skills and knowledge developed through the inquiry.

Skills

The PSTs discussed four skills when describing their experience of carrying out inquiry in the context of SSI: research, present information, propose investigatable questions, and self-management.

When talking about the secondary research they carried out as part of their inquiry, the PSTs often simply used the single word "research" and did not elaborate. One student elaborated by describing "finding relevant info among various sources; being able to sieve out the rubbish". The sources of evidence for their research were described as "internet" and "books". PSTs also talked about how to choose appropriate sources of information such as "how to distinguish between good and bad sources" and recognize "biased sources" and "neutral sources". They referred to finding sources of evidence that "supported" and "were against" the chosen research question or contained "different points of view".

The PSTs discussed developing the skills relating to presenting information by writing "coherent, structured reports", referring to "layout", "headings" and "word counts". The PSTs discussed proposing investigatable questions and stated changes they would make to their questions for investigation such as "think more carefully about the question, make it narrower so it is easier to answer in conclusion", "choose a better question, in order to find out more, and discuss the argument from both sides", "pick a question that gives me a little bit more room to research". The PSTs discussed self-management including managing their time by "organizing and sifting through information in a short period of time" or "spend more time" on some aspects of the inquiry. They also talked about self-organization such as "set out goals before I looked up information" and "do an initial plan before carrying out the report".

Knowledge

There were two sub-themes within the knowledge theme. Implications of scientific knowledge on society was the larger of the sub-themes compared to recall and application of scientific knowledge.

The PSTs demonstrated their knowledge of how their chosen technological application of physics impacts society, making statements such as "finding out what I wanted to know about the health effects of MRI scans: No major long term health effects." and "MRI scans can be harmful to unborn babies in the first 3 months of pregnancy as they can cause the human body to slightly heat up". The PSTs recognized limitations to how they situated their scientific knowledge within the SSI context, stating "[If I carried this out again I would] be more focused on how speed cameras affected no of deaths and less talk about how speed cameras operate".

The PSTs also demonstrated their knowledge of the science behind the technological applications, without relating it to the SSI context, stating "[I] develop[ed] understanding of what GPS is and how it works" and "I feel I gained a lot of new information on how the physics phenomenon of solar cells causes solar heat and solar electricity."

Discussion

Overview and SSI Context

At the beginning of phase one, the PSTs, with support from the facilitator, chose questions for investigation which were based on scientific knowledge and situated within an SSI context, specifically the societal implications of a chosen technological application of physics. They then began to carry out research, searching for information relevant to their research question from a range of sources. In the last phase of the task, carried out the following week, the PSTs communicated their findings as a written report. Throughout the process the PSTs demonstrated self-management in terms of time management and planning.

The SSI context of the task related to the societal implications of a technological application of physics. SSI are more than simply "real life" contexts; they are controversial, contemporary and relevant [8, 9]. They involve a number of conflicting scientific, social or moral viewpoints, which may conflict with the students' own views, this makes them personally relevant to the students [8, 9]. The SSI cannot be concluded definitely, even after thorough examination of evidence [8]. In this case study the PSTs had some choice in their question for investigation which was intended to allow the PSTs to choose a personally relevant issue. However, there is evidence that some of the contexts chosen by the PSTs lacked the range of controversial viewpoints required of a true SSI. In other cases, the SSI contexts were not explored in terms of the social, moral or ethical implications and instead the PSTs were overly focused on the science, e.g. how the technology works. In future iterations of this inquiry, criteria for choosing an SSI (e.g. controversial, contemporary, relevant) will be explicitly discussed with the PSTs. This may go some way to addressing the lack of "authentic" SSI explored by PSTs.

Skills and Knowledge of Science

The PSTs emphasized research. They discussed finding and evaluating a range of sources of evidence and viewpoints. Researching for information is different from searching for information. Where search involves looking for "information or facts", research involves "putting different pieces of information together to find patterns, correlations and connections" [10; 0.46–1.00 min]. Research involves using multiple resources and, crucially, thinking "critically about the information found" [10; 1.22–1.31 min]. There was evidence that the PSTs went beyond simply searching and engaged in research into the SSI context.

When discussing presenting their findings, PSTs mainly talked about the structure of their report. This was a relatively large focus for the PSTs. This skill was performed without reference to the SSI context. When discussing their chosen question for investigation, the PSTs tended to be critical. Some felt that a poorly chosen question limited the scope for exploration of the SSI by not allowing them to explore a range of viewpoints or being overly focused on the science behind the context. This indicates the importance of supporting students when choosing a question for investigation. It may be advisable for students to be given the opportunity to revise their question after initial research in a question/action cycle [11].

The PSTs discussed the knowledge acquired through the inquiry mainly within the context of the SSI explored. They described the implications of the technology on society and the environment. They also applied their scientific knowledge as they discussed the science behind the technology.

Implications

These findings can be discussed in terms of implications for the implementation of inquiry in the context of SSI in secondary schools. The PSTs' experience as learners is analogous to that of secondary school students, although the level of performance of skill and demonstration of knowledge would be expected to be higher. The teaching approach taken in this inquiry was student-led, research-based inquiry and this resulted in a number of skills being developed by the students with the main focus being research and presentation of findings. The knowledge developed was highly context specific. The choice of SSI context, careful question choice and opportunities to revisit and revise the question are important facets in allowing students to demonstrate their skills and knowledge in the context of the SSI. This study can also be discussed in terms of the implications for initial teacher education programs. The PSTs acted as learners with little focus on developing the skills relating to pedagogical approaches for facilitating inquiry. Future iterations of this module should allow the PSTs to explicitly reflect on their experience in terms of the teaching approaches used and how they could implement these in their own classrooms.

References

1. Sadler, T.D.: Situated learning in science education: socio-scientific issues as contexts for practice. Stud. Sci. Educ. **45**(1), 1–42 (2009). https://doi.org/10.1080/03057260802681839
2. Wee, B., Shepardson, D., Fast, J., Harbor, J.: Teaching and learning about inquiry: insights and challenges in professional development. J. Sci. Teach. Educ. **18**(1), 63–89 (2007). https://doi.org/10.1007/s10972-006-9031-6
3. Roehrig, G.H., Luft, J.A.: Constraints experienced by beginning secondary science teachers in implementing scientific inquiry lessons. Int. J. Sci. Educ. **26**(1), 3–24 (2004). https://doi.org/10.1080/0950069022000070261
4. Bencze, J.L., Sperling, E.R.: Student teachers as advocates for student-led research-informed socioscientific activism. Can. J. Sci. Math. Techn. Educ. **12**(1), 62–85 (2012). https://doi.org/10.1080/14926156.2012.649054
5. National Council for Curriculum and Assessment (NCCA): Junior cycle science: guidelines for the classroom-based assessments and assessment task (2016)
6. National Council for Curriculum and Assessment (NCCA): Junior cycle science: curriculum specification (2015)
7. Braun, V., Clarke, V.: Using thematic analysis in psychology. Qual. Res. Psychol. **3**(3), 77–101 (2006). https://doi.org/10.1191/1478088706qp063oa
8. Levinson, R.: Towards a theoretical framework for teaching controversial socio-scientific issues. Int. J. Sci. Educ. **28**(10), 1201–1224 (2006). https://doi.org/10.1080/09500690600560753
9. Zeidler, D.L., Sadler, T.D., Applebaum, S., Callahan, B.E.: Advancing reflective judgment through socioscientific issues. J. Res. Sci. Teach. **46**(1), 74–101 (2009). https://doi.org/10.1002/tea.20281
10. McMaster Libraries: Search vs research. Online video. https://www.youtube.com/watch?v=minZ0ABVqyk. Last accessed: 2018/11/26
11. Exploratorium: Fundamentals of inquiry facilitator's guide workshop III: raising questions. Exploratorium, USA (2006)

Responsibility of Teachers: The SSIBL Model in Hungary

Andrea Király⬤, Andrea Kárpáti⬤ and Péter Tasnádi⬤

Abstract Physics teacher education traditionally focuses on a science-grounded educational approach that involves experimentation in the form of teacher demonstrations and student laboratory work. Introducing the SSIBL model in the framework of an in-service training course, "Contemporary methods in Physics Education", required an in-depth introduction of the components of the model but also a case-based explanation of the relevance and necessity of socially sensitive science education for Hungary. The development of the course started with an immersion of teachers in science communication media related to burning social issues like the use and abuse of nuclear energy, the potentials of renewable energy, safe and conscientious energy consumption etc. Background knowledge gained during the course was reinforced by the 2015 international conference in Budapest, "Teaching Physics Innovatively". Many presentations, symposia and a roundtable discussion targeted components and philosophy behind the SSIBL model. One much discussed social issue, the necessity of the expansion of the Nuclear Plant at the town of Paks was also supported by a guided site visit preceded by a roundtable discussion and followed by further elaboration during the TPD course. A survey supplemented by interviews was used to identify results and problems of the introduction of SSIBL in Hungary. This paper provides an overview of the TPD course, the projects designed by the teachers and the general perception of the SSIBL model by Hungarian secondary school teachers of Physics.

Keywords Socially sensitive issues · Environmental issues · GLOBE program

A. Király · A. Kárpáti · P. Tasnádi (✉)
Faculty of Science, ELTE Eötvös Loránd University, Budapest, Hungary
e-mail: ttasipeter@gmail.com

A. Király
e-mail: andrea.kiraly@ttk.elte.hu

A. Kárpáti
e-mail: andrea.karpati@ttk.elte.hu

A. Király · P. Tasnádi
MTA-ELTE Physics Education Research Group, Budapest, Hungary

© Springer Nature Switzerland AG 2019
E. McLoughlin and P. van Kampen (eds.), *Concepts, Strategies and Models to Enhance Physics Teaching and Learning*,
https://doi.org/10.1007/978-3-030-18137-6_13

Introduction

Hungarian citizens often lack appropriate knowledge about developments in science and technology and therefore, the intensity and efficiency of science education has to be increased. Science teachers are instrumental in providing authentic and age-relevant information and sharing values and attitudes about the role of science in solving crucial social, economic or health related problems for students and their families. The results of science education, therefore, should not be restricted to targeting high performers and transmit knowledge and skills necessary to embark on a scientific or technological career. Our working group has initiated new methodologies through making science teachers aware of their role in teaching about socially relevant issues.

We joined the *Promoting Attainment of Responsible Research and Innovation in Science Education (PARRISE)* project, a Seventh Framework Program (Grant Agreement No. 612438, duration: 2014–2017) to expand our perspectives and enrich our repertoire of socially responsive teaching and education. Our in-service training program is based on the framework for *Socio-Scientific Inquiry-Based Learning (SSIBL)* [1] that is founded on Responsible Research and Innovation (RRI). Each of its three components: Socio-scientific Issues (SSI), Citizenship Education (CE) and IBSE (Inquiry Based Science Education) address challenges and risks of scientific research. This complex framework reflects that science education needs to address issues of social relevance and empower students to become responsible adults who are able and willing to influence political decisions influenced by scientific research.

The SSIBL Framework: A Social Engagement Model for Science Education

The SSIBL Framework was developed by a European community of science teachers, teacher educators and educational researchers who participated in the PARRISE project. One of the main objectives of this pedagogical model is to increase the agency and motivation of young people to pursue studies in science. The project objectives are as follows (cf. [2] for details and publications):

- Provide an overall educational framework for socio-scientific inquiry-based learning (SSIBL) in formal and informal learning environments;
- Identify examples of best practice;
- Build transnational communities consisting of science teachers, science teacher educators, science communicators, and curriculum and citizenship education experts to implement good practices of SSIBL;
- Develop the SSIBL competencies among European primary and secondary science teachers and teacher educators;

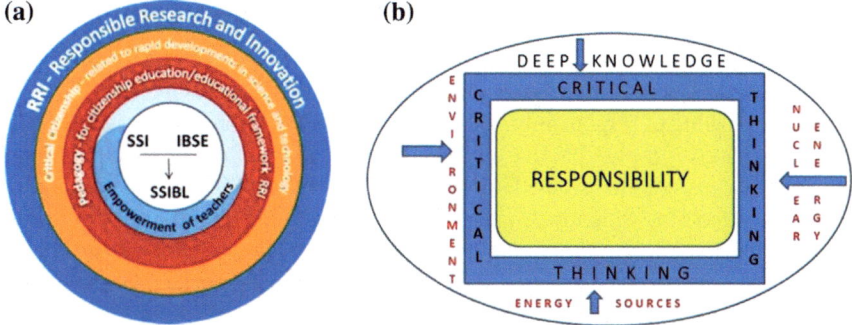

Fig. 1 **a** The Socio-scientific inquiry-based learning (SSIBL) model. *Source* [1]; **b** Hungarian model of socially-sensitive teaching

- Disseminate resources and best practice through PARRISE website, digital and print-based publications online and face to face courses authored by national and international networks;
- Evaluate the educators' success using the improved SSIBL materials with preservice and in-service teachers.

The project team collected and shared existing best practices in European science education and developed learning tools, materials and professional development courses for the SSIBL approach. The interrelations of the pillars of the model are represented in Fig. 1a.

The model is based on the concept of Responsible Research and Innovation (RRI) which is connected with three pedagogical concepts:

- *Inquiry based Science Education (IBSE)*: this model, a core component of Hungarian Physics education, is gradually being adapted by other science disciplines as well. It focuses on empowering students to act as researchers and offers them problems and solution scenarios to experiment with.
- *Socio-scientific Issues (SSI)* are open-ended science problems that often involve controversial social issues closely connected to research and innovation in science. SSI encourage teachers to apply both scientific and moral argumentation and develop solutions in relation to real-world situations like climate change, genetic engineering, or the use of nuclear power as cheap and clean energy resource.
- *Citizenship Education (CE)* is aimed at educating clear-thinking and enlightened citizens who participate in decisions concerning society, are aware of the rules of law and other regulations that concern interrelationships of social life and scientific innovation.

Adaptation of the SSIBL Framework for Science Education

As mentioned earlier, the main goal of education for social sensitivity is the development of the feeling of responsibility in students (Fig. 1b). Responsibility is a virtue of mind which makes it possible that students become adults who are capable of making socially sensitive decisions. However, motivation is not enough to make good decisions concerning questions connected with natural sciences. The prerequisite of responsible thinking is an adequate knowledge base of the issue discussed. In the case of scientific and technological innovations, we should be aware of the advantages and dangers of these. We should be able to estimate the risk of the technical tools. To make the right decision, we have to think critically and independently. We should avoid being manipulated by the press and other media. Furthermore, we should be aware of some simple facts about the working of the human brain and self-knowledge.

Hysteresis of Our Brain

Human beings are thinking in patterns and it is easy to demonstrate that our brain (or thinking), as a typical non-linear system with memory properties, exhibits the hysteresis phenomenon [3]. Therefore the environment of a problem can greatly influence its solution.

Figure 2 shows a series of eight small pictures. If we look at the pictures not at once but one after the other, it turns out that depending on whether they follow each other right-to-left or left-to-right, a different figure will be perceived on the drawing which is in the middle of the series. (In right-to-left order we see a female nude figure and in left-to-right order a cartoon man head). This visual phenomenon can be interpreted as a hysteresis of the processing of visual information in our brain as explained by the curve visible above the small pictures [4].

Another and maybe more impressive illustration of this phenomenon is a little joke which can be seen at [5]. It shows that the answer to an opinion poll can be manipulated if before the real question some other, misleading questions are asked. The question was whether the polled person supports the introduction of military service or not. The same person gives a completely opposing answer depending on the series of questions they were asked previously. The video is very convincing in spite of showing an imaginary (but plausible) situation.

Deep Knowledge

As mentioned above, to make a good decision about a debated question, we need deep knowledge of the concepts of science regarding the task discussed. In the following,

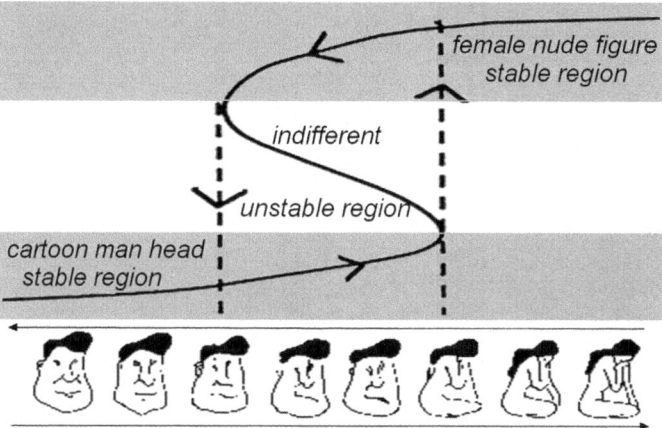

Fig. 2 Visual hysteresis. *Source* [4]

we are going to illustrate what we mean by deep knowledge through the discussion of a simple problem concerning velocity.

Let us solve the following problem: two escalators move in parallel. The average velocity of both escalators directs downward. Can a person who can arbitrarily change escalators but not use any other means move upward? Although at the first sight in seems to be impossible, the task can be solved.

Solution: Suppose that escalator (A) is moving uniformly with a speed v downward while escalator (B) ascends for a short time with speed $2v$ and then, for the same duration, moves downwards with speed $4v$. The average velocity of escalator B thus points downwards, and is equal to that of escalator A. It is obvious that the person standing continuously on either escalator will be descending, but when both of the escalators are moving at the same time, the person can ascend if he/she change escalators at appropriate times. The person should be on escalator B while it is ascending, and change to escalator A when it is descending. When escalator B begins to ascend again, the person should step back to escalator A. Repeating this procedure continuously, the person will be getting upwards on average. The example shows why it is important to make a clear distinction between the concept of instantaneous and average velocity. Figure 3 illustrates the solution, but more convincingly, it can be understood through the simulation in [6].

Antecedents of SSIBL Methods in Hungary

In our opinion, perhaps the most important prerequisite of the IBL in physics is the use of an inductive and experiment-based teaching method. In Hungary, there is a great tradition of this kind of teaching methodology. In some important fields

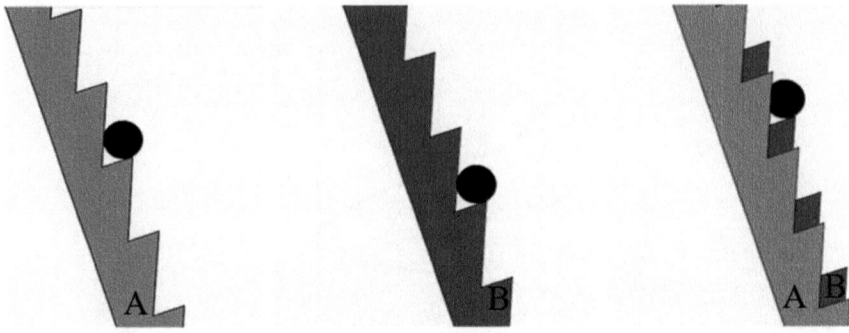

Fig. 3 Motion on the escalators. *Source* [6]

of physics (nuclear physics, environmental physics, sustainable development, risk estimation etc.), the association of social and scientific content has been realized. As we have mentioned before, the SSIBL has an inherent interdisciplinary character. In 1992, to promote interdisciplinary teaching in the field of environmental science at secondary schools, an in-service training course for Environmental Studies teachers was launched at the ELTE. In the following, this early appearance of a model similar to IBSE will be illustrated with some examples.

Measurements of Radioactivity

In Hungarian schools, both at elementary and secondary level, there has been a demand to connect scientific content of the curriculum to everyday life. This require-ment has been being strengthened with the development of technical tools. However, in addition to the obvious advantages of new inventions many have realized that some (nuclear energy, aviation, genetic engineering etc.) could be very dangerous too. Sometimes, for example in case of the use of nuclear energy, the danger has often been exaggerated. Teachers should have helped students to balance between the extreme views and to estimate the real hazards. George Marx, a Hungarian profes-sor of theoretical physics, pioneered the introduction of risk estimation in education. He published a three-part series of articles [7–9] on the risks of modern technologies and the importance of understanding the role of the estimation of risk. He wanted to show that "low risk" is acceptable, but the measure of low risk depends on our decision. For example, in the nuclear field, the risk is determined by the ALARA principle (As Low As Reasonably Achievable). He was particularly sensitive to mis-understandings concerning the use of nuclear energy. (The use of nuclear energy in our day is also a key question of the Hungarian economy [10].) In 1993, George Marx wrote a story [9] connected with the Chernobyl accident. The story took place in a Hungarian school. (In May 1986, Hungarian television announced that a radioactive cloud had reached Hungary.) Hungarian high school students at that time performed measurements with Geiger-Müller tubes; they measured also the background activity.

George Marx's story [9] about a Hungarian school. In the morning of the announcement, at a school, pupils queued at the door of the physics laboratory, early in the morning, waiting for the teacher. They wanted to measure the background activity again. It caused a great excitement when it turned out that the activity was three times higher than a month before. Students opened the windows to let the Chernobyl radioactivity in and measured the background again. Surprisingly, it fell to the old value. The morning increase was due to the accumulation of radon in the unventilated classroom during the night. The lesson was unforgettable for these students. As George Marx said: "Radioactivity around us is a fact of life. Nuclear fallout can be measured exactly, as we did in Hungarian secondary schools after Chernobyl, and as we have monitored radon since. High technology can be controlled. Understanding the facts influence the response of citizens, and thus the collective decisions of a nation in important questions. This is a prerequisite for the realization of democracy."

The GLOBE Program

The GLOBE (Global Learning and Observations to Benefit the Environment) program [11] was launched 1994 in the USA with the goal of promoting the teaching and learning of science, with giving an opportunity worldwide for students to participate in data collection and the scientific process to better understand, sustain, and improve Earth's environment. In the frame of an environmental observational system, students make measurements and observations in the field of meteorology, water chemistry, botany and zoology. Hungary joined to the GLOBE program in 1999. At present, 30 secondary schools are participating in the program with the financial support of the Ministry of Education [12].

Environmental education can be efficient only if it treats the world both in its entirety and its constituent parts. This is the reason why we organize yearly student competitions in photography, poetry, short story writing and drawing that demonstrate the relationship between man and its environment.

A University Course on Everyday Physics

Even though a main objective of physics is the explanation of everyday phenomena and processes, regular university courses are often too theoretical and do not concentrate on showing the application of basic laws in the interpretation of everyday phenomena. Such an approach would be particularly important for future teachers. However, for a teacher, it is not enough to show the role of physical laws through the explanation of everyday phenomena, they should also be able to interpret this explanation at secondary school level. Therefore, a course with title of Everyday Physics was built in the curriculum of pre- and in-service teacher education. Its content is changed according to the students' interest. The main topics generally involve the physics of cars (acceleration and braking, movement in the bend, physics of Formula 1, energy consumption of cars etc.), some questions about the weather (clouds, storms, lightning, weather forecast etc.) and the physics of sports.

Current Activities Related to SSIBL

Presently we are focusing on in-service teacher training. The main form of mediating IBSE and RRI toward teachers is an in-service teacher training course embedded in a Physics Education Ph.D. program.

The Physics Education Ph.D. program of the Graduate School for Physics was launched 10 years ago in Budapest at the Eötvös Loránd University, as a possible measure against the continuous decrease in interest in physics among high school students. The new program declares that establishing a novel, inspiring way of teaching some aspects of physics in a class is an achievement equivalent to traditional research results. The program is special since it is tailored specifically to the needs of active, in-service physics teachers, and the candidates carry out their research at their own school [13].

The research work is started right after entering the Ph.D. school, but the first two years are devoted mainly to lectures. The four semesters, equal to four modules, each consisting of four courses from different fields of physics, such as astronomy, modern physics, chaotic dynamics and cross-curricular aspects of physics and chemistry or biology, etc. But each module contains a course especially devoted to teaching methodology of physics.

The TPD of PARRISE

We developed our Teacher Professional Development course ("Current Contents and Methodology in Teaching Physics in the 21th Century") for in-service teachers, based on the four modules of the Physics Education Ph.D. program, customized according to the educational framework provided by PARRISE [2] colleagues and the national context. This enables science teachers to incorporate SSIBL. The course was officially accredited and nationally validated in September 2014. We finished the pilot course in the spring of 2015 with 20 participants, who, without exception, are practicing Physics teachers. During the course, they learnt about current results of Physics as well as the methodology of teaching Physics at primary and secondary school level. Methodology focused on the pillars of the PARRISE framework and included good practice presentations on IBSE, PBL, RRI and on the educational interpretation of social sensitive scientific issues.

One possibility for the completion of the TPD course is to write a pedagogical essay or to prepare e-learning materials on the topics of that module, e.g. on teaching strategies of classical physics, historically outstanding experiments in physics, relativity and cross-curricular aspects of physics and chemistry. Another option for course assessment is to take an oral examination or a written examination, or both, on the course material and earn university credit points that may be validated during doctoral studies in the Ph.D. program for Physics Education at the Graduate School for Physics of ELTE.

Case studies: The above requirements directly lead to the use of the elements of SSIBL at secondary schools [14]. The following case studies summarized below may serve as examples.

At the conference organised in 2015 by the Ph.D. school of ELTE in co-operation with the Hungarian PARRISE team, a series of lectures focused on good practice in physics education. A round table discussion was organised to feature different aspects of nuclear energy use [15] and the planned expansion of the nuclear power plant at Paks, a medium size Hungarian town. The plant was visited by conference participants who could witness the enhanced security measures and learn about the procedure of producing "clean energy" [16].

Ildikó Takáts Lucz developed a project on the use of renewable energy, the ecological footprint of man, and global environmental problems related to physics [17]. She employed project methodology also to engage 68 seventeen-year-old students in exploring these issues. Students worked in small groups and shared their results on the internet.

In the project by Edit Szombati, students measured the efficiency of light bulbs in use in Hungary [18]. There are four types of bulbs available: traditional Wolfram filaments, halogen, compact and LED light sources. Students performed measurements in the school lab with each of them, compared the relative light output and the heat radiation of the bulbs.

Balázs Kovács performed an educational experiment in two classes with 16 and 17-year-old students, 45 participants in total. Students received project tasks related to sparing water and electric energy. The projects involved measurements at home, to be performed individually, collection and comparison of data, calculations performed in class or in groups, and preparing presentations about the results.

Borbála Leitner organized informal learning projects outside the walls of the school building. She took her 13-year-old students on a field trip to see the water plant at the town called Tiszalök. The students took part in a guided tour where the functioning of the plant, the ways how energy is produced and transmitted, and the effects of floods were explained.

An Optional University Course

As a consequence of the GLOBE program, many students continued their secondary school education attending optional science courses. Therefore, our central goal is to integrate the GLOBE program more deeply into secondary level education. For integrating this program in regular secondary education, a one-semester optional course is offered for B.Sc and M.Sc students in earth and environmental sciences and for future teachers of different subjects (Physics, Biology, Geography, etc.). Besides environmental measurements, the course emphasizes the societal elements of sustainable development. An overview of the history of the predictions on the future of mankind and the earth is also provided from Malthus on, through the reports of the Club of Rome, till the recent UNESCO report "Shaping the Future We Want".

The Future

Research on integrating social issues in physics education continues in the framework of the MTA-ELTE Physics Education Research Group, e.g. [19]. We hope to develop projects that may show how this model may be utilised within the scope of the Hungarian Core Curriculum and share them with the physics education community.

Acknowledgements This work has received funding from the European Union's Seventh Framework Programme for research, technological development and demonstration under grant agreement No. 612438 (PARRISE, Promoting Attainment of Responsible Research and Innovation in Science Education) and from the Content Pedagogy Research Program of the Hungarian Academy of Sciences.

References

1. Levinson, R., PARRISE consortium: Socio-scientific based learning: taking off from STEP-WISE. In Bencze, J.L. (ed.) Science & Technology Education Promoting Wellbeing for Individuals, Societies Environments, pp. 477–502. Dordrecht: Springer (2017). https://www.parrise.eu/dissemination/#1475593741045-14cb400b-57d0, last accessed 2018/08/17
2. PARRISE (Promoting Attainment of Responsible Research and Innovation in Science Education) Project website: https://www.parrise.eu/about-parrise/, last accessed 2018/08/17
3. Haken, H.: Synergetics. An Introduction, 1st edn. Springer, Berlin. ISBN: 978-3-642-96365-0 (1977). https://doi.org/10.1007/978-3-642-96363-6
4. Nagy, P., Tasnádi, P.: Zeeman catastrophe machines as a toolkit for teaching chaos. Eur. J. Phys. **35**(1), 015018 (2014). https://doi.org/10.1088/0143-0807/35/1/015018
5. Mind manipulation—film. http://indavideo.hu/video/Mind_manipulation, last accessed 2018/08/17
6. Parrondo-paradoxon mozgólépcső (Hungarian). http://indavideo.hu/video/Parrondo-paradoxon_mozgolepcso, last accessed 2018/08/17

7. Marx, G.: Everyday risks. Phys. Educ. **28**(1), 23–25 (1993). https://doi.org/10.1088/0031-9120/28/1/004

8. Marx, G.: Risks of radioactivity. Phys. Educ. **28**(1), 121–125 (1993). https://doi.org/10.1088/0031-9120/28/2/011

9. Marx, G.: Public risks from the nuclear industry. Phys. Educ. **28**(1), 170–172 (1993). https://doi.org/10.1088/0031-9120/28/3/007

10. Aszódi, A.: Nuclear energy and public awareness. In: Király, A., Tél, T. (eds.) Teaching Physics Innovatively: New Learning Environments and Methods in Physics Education (e-book), pp. 393–401. Eötvös Loránd University, Budapest, ISBN 978-963-284-815-0 (2016). http://parrise.elte.hu/tpi-15/proceedings.php, last accessed 2018/08/17

11. The official website of the GLOBE (Global Learning and Observations to Benefit the Environment) program, https://www.globe.gov/, last accessed 2018/08/17

12. Weidinger, T., Kalapos, T., Gyuró, Gy., Dőry, I., Orgoványi, A., Victor, A.: The GLOBE program in Hungary. Geophysical Research Abstracts 8, A-08826 (2006)

13. Tél, T., Juhász, A.: Physics education Ph.D. program at Eötvös University, Budapest. Phys. Compet. **12**(2), 38–39 (2010). http://csodafizika.hu/fiztan/english/phd_eng.pdf, last accessed 2018/08/17

14. Király, A., Kárpáti, A., Tasnádi, P.: The past and presence of the SSIBL in Hungary. In: PARRISE Final Conference—Science and Society in Education—Promoting Responsible Research and Innovation through Science Education, 2017.08.20, Dublin, Ireland—poster. https://www.parrise.eu/parrise-final-conference-posters/, last accessed 2018/08/17

15. Aszódi, A., Featonby, D., Fülöp, Z., Salmi, H., Egyed, L.: Roundtable discussion about socially sensitive issues in physics education. In: Király, A., Tél, T. (eds.) Teaching Physics Innovatively: New Learning Environments and Methods in Physics Education (e-book), pp. 423–428. Eötvös Loránd University, Budapest (2016)

16. Náfrádi, G.: Visit to the maintenance and training center at Paks Nuclear Power Plant. In: Király, A., Tél, T. (eds.) Teaching Physics Innovatively: New Learning Environments and Methods in Physics Education (e-book), pp. 415–420. Eötvös Loránd University, Budapest (2016)

17. Takátsné Lucz, I.: Renewable sources in use—poster. In: PARRISE Final Conference—Science and Society in Education—Promoting Responsible Research and Innovation through Science Education, 2017.08.20, Dublin, Ireland—poster (2017)

18. Szombati, E.: Light bulb investigation—poster. In: PARRISE Final Conference—Science and Society in Education—Promoting Responsible Research and Innovation through Science Education, 2017.08.20, Dublin, Ireland—poster (2017)

19. MTA-ELTE Physics Education Research Group—Science Centres and Informal Learning Working Group. http://scilwg.elte.hu/, last accessed 2018/08/17

Using Self-video-based Discourse in Training Physics Teachers

Yaron Lehavi⦾, Avraham Merzel⦾, Ruti Segal, Ami Baram⦾
and Bat-Sheva Eylon⦾

Abstract Videotaping is widely accepted as a useful tool for teachers' education and learning. Analyzing their own teaching videos was found to be especially valuable for those teachers who became aware of more relevant components of teaching and learning and induced changes in a wide range of their pedagogical habits. From another perspective, educational reformers pointed out that increasing teachers' engagement in peer discourse is linked to greater student achievements, an innovative climate and improved educational reform implementation and sustainability. Here we describe a study of an approach designed to develop physics and mathematics educators' proficiency in conducting a peer video based didactic discourse. Our preliminary results indicate that participation in training workshops according to this approach enhanced the participants' level of awareness [1] and abilities to focus their attention on subject matter (physics and mathematics and their teaching) oriented issues.

Keywords Professional development · Teachers' instructors · Self-videoing · Levels of awareness

Y. Lehavi (✉)
The David Yellin Academic College of Education, Jerusalem, Israel
e-mail: yarlehavi@gmail.com

Y. Lehavi · R. Segal · A. Baram · B.-S. Eylon
The Weizmann Institute of Science, Rehovot, Israel
e-mail: rutisegal@gmail.com

A. Baram
e-mail: ami.baram@gmail.com

B.-S. Eylon
e-mail: bat-sheva.eylon@weizmann.ac.il

A. Merzel
The Hebrew University of Jerusalem, Jerusalem, Israel
e-mail: alef.mem@gmail.com

R. Segal
Oranim Academic College of Education, Kiryat Tiv'on, Israel

© Springer Nature Switzerland AG 2019
E. McLoughlin and P. van Kampen (eds.), *Concepts, Strategies and Models to Enhance Physics Teaching and Learning*,
https://doi.org/10.1007/978-3-030-18137-6_14

159

Introduction

Videotaping is widely accepted as a useful tool for teachers' education and learning [2–12]. Teachers are commonly encouraged to observe either their own teaching videos or those of others [13–15]. It was found that, compared to analyzing other teachers' videos, teachers who analyzed their own teaching experienced higher activation, manifested by their higher immersion and motivation, and they became aware of more components which are important for teaching and learning. Moreover, the use of video encouraged changes in teachers' pedagogical habits because it helped them to focus on analyzing their choices, view their teaching from a new perspective, trust the feedback they received, feel free to change their practice, remember to implement changes, and see their progress [16]. However, they were also less self-reflective with regard to articulating critical incidents [17]. From another perspective, educational reformers have advocated for increasing teachers' engagement with peer observation, feedback, and support [18], since such collaboration have been linked to greater student achievement, an innovative climate, as well as improved reform implementation and sustainability [19–22].

The widely agreed potential of using video analysis in teachers' professional development (PD) has ushered in the development of a special video-based didactics for teachers' training. This led us to develop a Video-Based Didactics (VBD) program whose aim is to provide math and physics teachers with professional development by using video analysis effectively, both at pre-service and in-service levels as well as for teachers-trainers' PD. The program utilizes an approach that emphasizes a specially designed Video-Based Didactic discourse (abbreviated hereafter as VBD discourse) generated purely for the teachers' PD and not for their administrative evaluation. The approach is based on the assumption that teachers' privacy and independence are highly important factors in teaching as a profession. It also takes into account the fact that although video analysis offers a great opportunity for teachers' PD, not all of them are comfortable in opening themselves up to such scrutiny [15].

Here we will describe a study aimed at revealing how using our VBD program changed the focus of attention from a general pedagogy focus to physics and physics teaching focus at different levels of physics education ranging from teachers to physics teachers' trainers.

Theoretical Framework

Teachers' professional growth involves climbing from lower to higher levels of awareness with regard to teaching. Here we adopted Mason's framework [1] which involves three levels of awareness:

- Level 1: Awareness in action: awareness of the ability to make didactical choices with regard to teaching physics or mathematics.

- Level 2: Awareness in discipline (knowledge of awareness in action): awareness of the ability to examine how the teacher performs the actions mentioned in the previous step, while addressing the knowledge of the discipline. For example, physics teachers who are aware of the Phys-Math interrelations while fostering their students' mathematical/physical thinking habits exhibit level 2 awareness. Mastering this level indicates that the teachers know the importance of being aware of their actions in the classroom, and their consequences.
- Level 3: Awareness in counsel (knowledge of awareness in discipline): At this level, self-awareness is required to become sensitive to the needs of others in order to build one's own awareness in actions and awareness in disciplines. Mastering this level may indicate professionalism on behalf of the teachers as mentors of other teachers in their own discipline.

In the next section we will describe the VBD approach and will show some evidence that its implementation induces awareness at all three levels.

Principles of the VBD Approach

Our VBD approach was developed in view of the previously described growing use of video analysis in teachers' PD, both preservice and in-service, and the corresponding need for suitable special didactics. Our VBD approach is based on the following rules [23, 24]:

1. **Use of evidence**—The VBD discourse builds on evidence (excerpts of video recordings) from the teacher's class or a mentor-mentee meeting. Both the mentor and the mentee are encouraged to use observations in order to formulate questions about their actual actions when playing the two roles (the mentor and the mentee).
2. **Ownership**—The teachers use their own device (commonly their own smartphone) to record their teaching and thus own it.
3. **Autonomy**—The teachers choose a short episode (5–7 min-long) from a whole lesson, on which the VBD discourse will be based.
4. **Distinct role**—In every VBD discourse the teacher who brings the video evidence is the mentee and his partner is the mentor. Thus, even if the mutual conversation is held between colleagues of the same level of professionalism, the role played by each of them is clearly defined. They may switch roles from one session to another but in each session the roles are clearly defined.
5. **Shared professionalism**—The subject matter and professional expertise should be shared by the mentor and the mentee, in order to enable a thorough content-related discourse. Thus, they both should have experience in teaching the same subject matter.
6. **Mutual development**—The mentor and the mentee develop together their analysis tools. The development process is mutual, even if one brings to it a richer set of such tools than the other.

7. **Introspection**—The process of choosing an episode from a whole lesson, and provide it with a header, initiates an introspection process both at the cognitive and the affective dimensions.
8. **An inquiry focus**—Similar to the constructivist approach for teaching [6 for example], discovery plays a central role in the VBD discourse and the mentor and the mentee who inspect a video recording learn how to analyze, conceptualize, and evaluate the scene appearing in it. They are encouraged to adopt a curious eye rather than a judgmental one when looking at the segment for topics to be discussed during the discourse. The discourse itself is carried out as a conversation between the trainer and the trainee who share their interpretations of the video evidence at hand.

The VBD Professional Development Program

Based on the principles mentioned above (mainly 6–8), we developed a professional development program for physics (and mathematics) educators in order to promote their VBD skills when engaged as mentors in a VBD discourse. The VBD principles were conveyed to physics and mathematics educators (separately) through specially designed workshops in which the attendees participated in a VBD discourse both as mentors and as mentees.

The program followed three steps: (a) The mentee chooses a short episode (5–7 min) from the videotape of his/her own lesson and gives it a title. (b) The mentor receives the video segment and prepares himself/herself for the didactic discourse by watching it (usually more than once) with a critical, yet not a judgmental, eye and formulates accordingly inquiring questions for the discourse. We encouraged our participants to choose questions that deal with the subject matter or subject matter teaching, by looking for evidences for students' possible difficulties in understanding what is taught. (c) The mentor and the mentee meet (face to face or online) and discuss the inquiries raised by both of them. The meeting is recorded and becomes the common property of them both. In many occasions the participants went to certain instants in the video as references for their discussion. The mutual inquiry-based discourse focuses on what the participants consider as interesting teaching aspects that were overlooked during class such as students' misconceptions and how they were addressed by the teacher's explanations and responses. (d) The mentors can use the videotaped meeting for a discussion with another mentor (a colleague or teachers' trainer or a VBD trainer) in order to improve his guidance skills.

The Research

The research questions. This is an ongoing study which (among other aspects) aims at characterizing the VBD discourse and evaluating its effectiveness for teachers and teachers-as-trainers. We therefore asked the following questions:

What support is provided by the VBD discourse to the teachers in their PD? What support is provided by the VBD discourse to the PD of teachers-as-trainers? What characterizes an effective VBD discourse?

Here we will share some of our conclusions with regard to the first two questions and some preliminary insights with regard to the last one. We adopted here Chapman's definition of an effective video-based discourse as one that generates new insights of new teaching ideas, strengthens one's awareness regarding the teaching-learning process, and creates critical incidents and significant turning points [25].

We looked for evidence regarding how the VBD discourse changes the mentor's and the mentee's level of awareness (according to Mason's framework) and focus their attention on teaching-learning instants that indicate on their awareness in action and awareness in discipline.

Methods

Target populations and research tools. We administered a questionnaire to the physics teachers in our workshops (N = 28; average age ~41 years, with average teaching experience of less than 5 years), requesting their views with regard to the workshop and a possible use of the VBD approach.

We divided our population into three groups according to their official role: (T^1) school teachers (N = 16) who participated in physics teachers' communities',[1] half of them (8) with teaching experience of more than 5 years; (T^2) Teachers that serve as teachers' instructors (N = 8); and ($T^{1.5}$) Teachers that serve as leaders of physics teachers' communities (N = 4). These teachers are not defined officially as teachers' instructors but sometimes are considered so by the teachers in their community.

The 8 relatively inexperienced physics teachers in group T^1 experienced a hierarchical VBD discourse with a teacher from group T^2. Apart from them, all the other teachers in this group (T^1), as well as the teachers from the T^2 and $T^{1.5}$ groups had participated in special workshops that demonstrated and practiced the VBD non-judgmental and inquiring discourse approach.

---

[1] The physics teachers' communities program was initiated by the Department of Science Teaching at the Weizmann Institute, Rehovot, Israel, and the department has operated it since 2012.

Table 1 The participants' views and beliefs regarding the VBD approach

	VBD usefulness in general	Continue using VBD as trainees	The value of using VBD as trainers		Usefulness of VBD for teachers' guidance
			Non-hierarchical	Hierarchical	
Group T^1 N = 16	M = 4.50 (sd = 1.26)	M = 3.75 (sd = 1.34)	N = 8; M = 3.63 (sd = 1.60)		
Group $T^{1.5}$ N = 4	M = 5.0 (sd = 0.82)	M = 5.0 (sd = 0.82)	M = 3.25 (sd = 0.96)	M = 4.0 (sd = 1.41)	
Group T^2 N = 8	M = 4.88 (sd = 0.99)	M = 4.38 (sd = 1.51)		M = 4.38 (sd = 1.41)	M = 4.88 (sd = 0.83)
Total	M = 4.68 (N = 28) (sd = 1.12)	M = 4.11 (N = 28) (sd = 1.37)	M = 3.50 (N = 12) (sd = 1.38)	M = 4.25 (N = 12) (sd = 1.36)	M = 4.88 (N = 8) (sd = 0.83)

Findings and Initial Interpretations

Views and beliefs.[2] We asked the participants for their opinions regarding the usefulness of the VBD in improving their teaching. The findings are summarized in Table 1.

Table 1 shows that generally the teachers' possessed positive views regarding the VBD (Overall: M = 4.68, sd = 1.12) and the possibility of using it in the future as mentors (M = 4.11, sd = 1.37). Those who played the mentor role (N = 24[3]) tended to be positive about using VBD (M = 3.88, sd = 1.39). However, the responses of teachers from groups $T^{1.5}$ and T^2 who participated in a hierarchical training were more enthusiastic (M = 4.25, N = 12, sd = 1.36) than of those from groups T^1 and $T^{1.5}$ who experienced only peers' VBD discourse (M = 3.50, N = 12, sd = 1.38). This difference might be due to the scant experience that T^1 teachers have in teaching (most of them, as mentioned before, had less than 5 years of classroom experience), or the scant experience of the teachers as trainers (with or without VBD). The last inference might be strengthened by the responses from the teachers in group T^2 who had ample experience in teachers' guidance. These teachers indicated that VBD could be very useful for them not only in their teaching (M = 4.88, N = 8, sd = 0.83) but also in providing guidance for other teachers (M = 4.88, N = 8, sd = 0.99).

Effectiveness of the VBD discourse. In the questionnaires, we asked the teachers to describe what they had learned through their experience with the VBD discourse about teaching (in general) and what they had learned about the subject matter (physics and teaching physics). We also asked them to indicate the part of the VBD discourse in which an important (to them) issue arose—whether it was during their

[2]The Likert-type scale employed here ranges from 1 (not at all) to 6 (very much).

[3]Eight of the 16 T^1 teachers experienced VBD as mentors during the workshop they attended. The $T^{1.5}$ teachers responded twice: for hierarchical and non-hierarchical training.

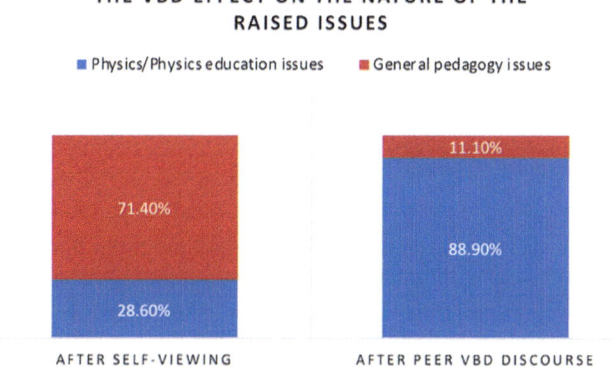

Fig. 1 Characteristics of issues emerged following the VBD discourse

self-viewing or within the interactions with others (as mentors or as mentees). There were 21 responses that explicitly mentioned such issues that arose following the self-viewing, of which only 6 (28.6%) responses related to the subject matter (physics and physics teaching); the rest focused on general pedagogy or teachers' behavior in class.

Examples of responses that mentioned issues that arose following the self-viewing (we used pseudonyms throughout the paper):

- **Reuven (T[1])**: "I've learned that I am too monotonous and I should change my voice intensity ... I learned that mainly when I watched myself".
- **Dina (T[1])**: "Watching myself, I've learned that I run around a lot in class".
- **Rachel (T[1])**: "...Sometimes, [my] classes seemed 'better' in the video than I thought [they were]. It gave me a better indication of the noise in class and the number of students who participated in the lesson ...".
- **Shimeon (T[1])**: "Watching [my] videos and getting to feel how the lesson looks from the rear bench, helped me to understand that I should be more assertive in maintaining discipline in my class".

There were 27 responses that explicitly mentioned issues that emerged following interactions with others; of these, 24 (88.9%) were related to the subject matter (physics or physics teaching). See Fig. 1.

Here are some examples provided by our addressees:

- **Levi (T[1])**: "While watching my colleague's video, I learned a way to teach the law of inertia using the example of a rock that falls from a ship masthead...".
- **Sara (T[2])**: "When the conversation with the mentor was about to end, I became aware of the importance of explaining "explosion" by using momentum conservation [which I did spontaneously] rather than by Newton's third law ...".
- **Lea (T[1])**: "The guidance helped me understand more clearly the situation of giving a student an incomplete answer. I would like to spend more time on the students' contributions to the lesson".

- **Yehuda (T^2):** [during a guided VBD discourse]: "Could adding energy considerations to calculating speeds in a horizontal throw be a proper strategy for teaching 9th grade students?"

All these examples indicate that the teachers' level of awareness in action (level 1 in Mason's framework) increased.

Twenty-eight responses did not indicate explicitly the part of the video in which the issues came up. Twelve (42.9%) were related to the subject matter (physics and physics teaching), 13 (46.4%) were related to pedagogy or teachers' actions, and 3 (10.7%) were concerned with other issues. No participants claimed to have learned nothing from the VBD discourse.

There were also a few responses that pointed out their insights as mentors and expressed a growing "counsel awareness" (in Mason's terminology). For example:

Dan (T^1): "Not always are we aware of all the implicit premises in the way we teach. That point became clear to me through my experience both as a mentee and as a mentor".

Rivka ($T^{1.5}$): "I'm very interested in 'how do we know that the students understand the material'. […] And how do we check this? This issue emerged in several guidance discourses [I made as a mentor] and while watching [colleagues'] videoed episodes".

Yehuda (T^2): "Instructing [other teachers] is not intuitive. It takes time and planning. It came up over and over again during the [VBD] program".

Rina (T^2): "Comparing video excerpts of different teachers that teach the same subject could be interesting."

These responses (except from 1 teacher, Rivka) are from teachers and teachers' instructors that had experienced the VBD discourse as a tool only several times (no more than two times as a mentee, and once (if at all), as a mentor).

Discussion

The VBD approach is aimed at helping physics and mathematics educators (teachers and teachers' instructors) in exploring and solidifying their self-views as professionals. We believe that in this respect it may evoke another level of awareness—awareness in council.

We have found that physics teachers of different levels (T^1, $T^{1.5}$, T^2) find the VBD discourse to be useful for improving their teaching and they expressed their willingness to continue and use this tool for further PD as physics teachers. These results suggest that VBD addresses the teachers' needs and those of their instructors for a PD tool in which one can delve deep into the taught subject matter in order to provide the teachers with meaningful insights about physics and physics teaching.

We have found that taking the mentor role is more convenient in a hierarchical VBD. In this respect, the scant training experience that T^1 teachers had, may lead some of them to interpret their peers' role in a discourse as a "feedback discourse", although this is not the purpose of our VBD approach. Possibly it was difficult for them to distinguish between an "inquiry observation", which looks for interesting instances in a lesson and what can be learned from discussing them (the core of what the VBD discourse is about) and a "judgmental observation", which examines a lesson through the right-wrong prism. We have some indications that if T^1 teachers will experience VBD more in the context of peer training or colleagues' guidance, with the aid of specially trained VBD advisors, they will better understand the nature of the discourse and will consider, similar to T^2 teachers, the VBD approach to be useful for their teaching. Apparently, our findings indicate that T^2 teachers, and perhaps even $T^{1.5}$ teachers, recognize that the mentor can benefit from the discourse just as much as the mentee (perhaps even more). Hence, they tended to take the mentor's role more willingly.

One might think that watching oneself teaching is sufficient for a professional development. Our findings indicate that while watching oneself teaching or a spontaneous discourse mainly give rise to teachers' consideration of behavioral and general pedagogical issues, a purposeful VBD discourse seems to have the advantage of focusing the attention also on subject matter and subject matter teaching issues. Apparently, the interaction with another teacher—watching a colleague's excerpt of a lesson and participating in a VBD discourse, gives rise to many discipline and discipline teaching issues. Such an engagement in the VBD discourse was found to be very meaningful for the physics teachers, lead them to experience what Chapman [25] coined as turning points in their awareness of their teaching. We also found that VBD has a positive influence on T^2 teachers, since they referred to their role as trainers and as teachers' instructors. As Yehuda said: "instructing teachers is not intuitive ... it takes time and planning". VBD could serve as a proper tool for developing guidance skills at all teachers' levels. For example, T^2 teachers can videotape a guidance discourse (either video-based or not) with T^1 teachers and conduct a VBD discourse with a T^2 teacher or even with a T^3 teacher. We found that the VBD discourse and the VBD training program could raise teachers' awareness with regard to their teaching according to Mason's levels 1 and 2 [1] (awareness in action and in discipline) for all teachers' groups, as well as teachers' trainers' awareness of their counsel (level 3) and by this means, could foster professional growth.

Plans for Future Research

The VBD approach has the potential to become a useful tool for teachers' trainers, leaders of teachers' communities or teachers' instructors. It could focus their guidance and create critical incidents and significant turning points in their awareness in council. However, using the VBD approach in our educational system, has not yet reached full maturation and we continue to implement the program in physics and mathematics teachers' communities as well as in groups of teachers' instructors.

We have presented here some preliminary results of an ongoing study that is concurrent with the VBD implementation program. This research continues by examining VBD discourses of peer and hierarchical T^1, $T^{1.5}$, T^2 and even trainers of teachers' educators (T^3 teachers). We are currently analyzing responses to questionnaires, recorded VBD meetings, interviews, and case studies among others. This research addresses the VBD discourse with regard to the characteristics of the turning points that follow it, its value for teachers (both as mentors or mentees) in fostering a better understanding of topics in physics or mathematics, and for adopting a more attentive approach to students' difficulties in understanding the subject matter. Furthermore, we are also interested in examining the effect of using the VBD approach in pre-service teacher education.

References

1. Mason, J.: Enabling teachers to be real teachers: necessary levels of awareness and structure of attention. J. Math. Teach. Educ. **1**(3), 243–267 (1998). https://doi.org/10.1023/A:1009973717476
2. Sherin, M.G.: New perspectives on the role of video in teacher education. In: Brophy, J. (ed.) Using Video in Teacher Education, pp. 1–28). Elsevier, Amsterdam (2004)
3. Santagata, R., Guarino, J.: Using video to teach future teachers to learn from teaching. ZDM Int. J. Math. Educ. **43**(1), 133–145 (2011). https://doi.org/10.1007/s11858-010-0292-3
4. Blomberg, G., Sherin, M.G., Renkl, A., Glogger, I., Seidel, T.: Understanding video as a tool for teacher education: investigating instructional strategies to promote reflection. Instr. Sci. **42**(3), 443–463 (2014). https://doi.org/10.1007/s11251-013-9281-6
5. Calandra, B., Rich, P.J.: Digital video for teacher education: research and practice. Routledge, New York (2015)
6. Honebein, P.C.: Seven goals for the design of constructivist learning environments. In: Wilson, B.G. (ed.) Constructivist learning environments: Case Studies in instructional design. Educational Technology Publications, Englewood Cliffs, pp. 11–24 (1996)
7. Sherin, M.G.: The development of teachers' professional vision in video clubs. In Goldman, R., Pea, R., Barron, B., Derry, S.J. (eds.) Video research in the learning sciences (pp. 383–395). Lawrence Erlbaum, Mahwah, NJ (2007)
8. van Es, E., Tunney, J., Seago, N., Goldsmith, L.T.: Facilitation practices for supporting teacher learning with video. In: Calandra, B., Rich, P.J. (eds.) Digital Video for Teacher Education: Research and Practice. Routledge, New York (2015)
9. van Es, E.: Participants' roles in the context of a video club. J. Learn. Sci. **18**(1), 100–137 (2009). https://doi.org/10.1080/10508400802581668
10. van Es, E., Sherin, M.G.: Mathematics teachers' "learning to notice" in the context of a video club. Teach. Teach. Educ. **24**(2), 244–276 (2008). https://doi.org/10.1016/j.tate.2006.11.005
11. Zhang, M., Koehler, M.J., Lundeberg, M.: Affordances and challenges of different types of video for teachers' professional development. In: Calandra, B., Rich, P.J. (eds.) Digital Video for Teacher Education: Research and Practice (Chap. 9). Routledge, New York (2015)
12. Sherin, M.G., van Es, E.: Effects of video club participation on teachers' professional vision. J. Teach. Educ. **60**(1), 20–37 (2009). https://doi.org/10.1177/0022487108328155
13. Borko, H., Jacobs, J., Eiteljorg, E., Pittman, M.: Video as a tool for fostering productive discussions in mathematics professional development. Teach. Teach. Educ. **24**(2), 417–436 (2008). https://doi.org/10.1016/j.tate.2006.11.012
14. Krammer, K., Ratzka, N., Klieme, E., Lipowsky, F., Pauli, C., Reusser, K.: Learning with classroom videos: conception and first results of an online teacher-training program. Zentralblatt für Didaktik der Mathematik **38**(5), 422–432 (2006). https://doi.org/10.1007/BF02652803
15. Sherin, M.G., Han, S.Y.: Teacher learning in the context of a video club. Teach. Teach. Educ. **20**(2), 163–183 (2004). https://doi.org/10.1016/j.tate.2003.08.001

16. Tripp, T.R., Rich, P.J.: The influence of video analysis on the process of teacher change. Teach. Teach. Educ. **28**(5), 728–739 (2012)
17. Seidel, T., Stürmer, K., Blomberg, G., Kobarg, M., Schwindt, K.: Teacher learning from analysis of videotaped classroom situations: does it make a difference whether teachers observe their own teaching or that of others? Teach. Teach. Educ. **27**(2), 259–267 (2011). https://doi.org/10.1016/j.tate.2010.08.009
18. Supovitz, A.J.: Developing communities of instructional practice. Teach. Coll. Rec. **104**, 1591–1626 (2002). https://doi.org/10.1111/1467-9620.00214
19. Borko, H.: Professional development and teacher learning: mapping the terrain. Educ. Res. **33**(8), 3–15 (2004). https://doi.org/10.3102/0013189X033008003
20. Leana, C., Pil, F.K.: Social capital and organizational performance: evidence from urban public schools. Organ. Sci. **17**, 353–366 (2006). https://doi.org/10.1287/orsc.1060.0191
21. Pil, F.K., Leana, C.: Applying organizational research to public school reform: the effects of teacher human and social capital on student performance. Acad. Manag. J. **52**(6), 1101–1124 (2009)
22. Sun, M., Wilhelm, A.G., Larson, C.J., Frank, K.A.: Exploring colleagues' professional influence on mathematics teachers' learning. Teach. Coll. Rec. **116**(6), 1–30 (2014)
23. Segal, R., Lehavi, Y., Merzel, A., Baram, A., Eylon, B-S.: Using self-video-based conversations in training mathematics teacher instructors. In: Bergqvist, E., Österholm, M., Granberg, C., Sumpter, L. (Eds.) Proceedings of the 42nd Conference of the International Group for the Psychology of Mathematics Education, vol. 1, pp. XX–YY. PME, Umeå (2018)
24. Lehavi, Y., Merzel, A., Segal, R., Baram, A., Eylon, B-S.: Using self-video-based conversations In training physics teachers. In: Talk presented at the Conference of International Research Group on Physics Teaching (GIREP), Dublin, Ireland (2017)
25. Chapman, O.: Mathematics teachers' perspectives of turning points in their teaching. In: Kaur, B., Ho, W.K., Toh, T.L., Choy, B.H. (eds.) Proceedings of the 41st Conference of the International Group for the Psychology of Mathematics Education, vol. 1. PME, Singapore (2017)

A Workshop Approach to Pre-service Physics Teacher Education

Paul van Kampen ⓘ

Abstract In many countries science teachers are expected to teach students both scientific knowledge and competencies. The importance attached to the mastery of scientific practices waxes and wanes, but for the last thirty years or so many government policy documents have emphasized the importance of inquiry-based science education as a way to teach scientific practices. In theory, inquiry is the way scientists work and the way students learn to think and act scientifically. In the classroom, all too often "inquiry" is essentially a relabeling of tightly scripted teaching focused on the acquisition of facts and the execution of step-by-step procedures and algorithms. This state of affairs poses serious challenges to pre-service science teachers (PSTs) and their educators. We know that teachers tend to teach the way they were taught; therefore teacher education programmes should broaden teachers' views of what is possible and desirable in the classroom. This chapter discusses a study of PSTs enrolled in a second-year undergraduate module that forms part of a concurrent four-year Bachelor's degree programme, in which they simultaneously study education, mathematics, physics, and chemistry. A workshop approach in which PSTs first carry out and then critique inquiry activities was developed that allowed the PSTs to broaden their views of science teaching, while at the same time allowing researchers to investigate their views of science teaching.

Keywords Physics teacher education · Inquiry · Scientific practices

Introduction

This study investigates the views of teaching held by pre-service science teachers (PSTs); to be precise, undergraduate students who have chosen to become science and mathematics teachers of students aged between 12 and 18 years. A module that

P. van Kampen (✉)
School of Physical Sciences & Centre for the Advancement of STEM Teaching and Learning,
Dublin City University, Dublin, Ireland
e-mail: Paul.van.Kampen@dcu.ie

© Springer Nature Switzerland AG 2019   171
E. McLoughlin and P. van Kampen (eds.), *Concepts, Strategies and Models to Enhance Physics Teaching and Learning*,
https://doi.org/10.1007/978-3-030-18137-6_15

focuses on the development of pedagogical content knowledge of the PSTs which they take early on in their degree programme serves as the vehicle for the study.

One of the key premises underpinning the module design is that PSTs need to broaden their vision of what good science teaching entails. In this chapter, I will illustrate the need for multiple perspectives on teaching by discussing three possible ways of teaching acceleration, and their affordances and constraints. I will then illustrate a workshop approach [1] we have adopted to help students develop these multiple perspectives, and discuss how the approach allowed the researcher to establish the PSTs' views of teaching physics as a process of inquiry at an early stage in their teacher education.

Affordances of Different Forms of Instruction

Acceleration Three Ways

To illustrate what different forms of instruction have to offer, consider three ways in which students may learn about acceleration.

Didactic Teaching. Traditional didactic teaching, or "exposition and practice", has gotten a lot of bad press, but it is easy to overlook how much craft knowledge is involved in deciding what to teach and how. In the case of teaching acceleration, do you rely on geometry, algebra, or calculus? Do you start with the general definition of acceleration, with average acceleration in one-dimension, or a real-life example? Good teachers, no matter how they teach, will probe their students' prior knowledge, know where students may struggle, know what lies ahead in the course; and their students matter to them. Good teachers who use a didactic approach choose to and are able to explain the topic clearly and in an interesting way.

To establish to what extent students had developed conceptual understanding of acceleration, Shaffer and McDermott [2] asked them to draw acceleration vectors for a car moving with constant speed along an oval track (Fig. 1). The question may be considered conceptual because it is qualitative, unfamiliar, and not routine to the students. Of a huge cohort of about 6900 students, taught by many different instructors in different years, only 20% correctly drew an acceleration vector perpendicular to the path; 15% drew vectors towards the center or the foci of the track; 10% drew tangent lines; and 20% indicated the acceleration was zero. By rounding these numbers to the nearest 5%, these percentages applied to all student cohorts after a lecture on the topic, independent of the instructor.

There is a large body of research that shows similar results for conceptual questions on all kinds of introductory physics topics [3, 4]. Two inescapable conclusions are to be drawn from this: (1) lecture instruction does not help most students develop conceptual understanding, and (2) because this observation does not depend on "how good" the lecture is deemed to be, it is likely to be due to the mode of instruction.

Fig. 1 Setting used to ascertain conceptual understanding of acceleration. Redrawn after Fig. 3(b) in [2]

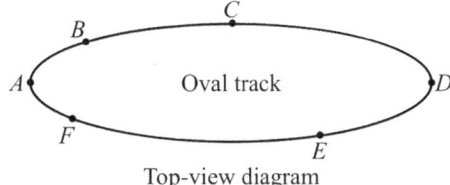

Top-view diagram

Guided Inquiry. It is often possible to help students develop conceptual understanding by creating a tutorial setting in which they discuss, in small groups, a carefully constructed teaching-learning sequence (TLS). These structured or guided inquiry approaches are often presented in the form of worksheets that students work through in a small-group tutorial setting. The instructor's role is not to provide answers, but to guide students towards a desired outcome. The TLS often achieves this by deepening students' operational understanding—in the case of acceleration, for example, they may construct change-in-velocity vectors and explore how they change with decreasing time intervals [1]. When such small-group tutorials by guided inquiry replace traditional tutorials revolving around end-of-chapter exercises, students' conceptual understanding may be brought to the level of graduate students while still maintaining their performance on standard examination questions [3].

Open Inquiry. It may appear that by introducing guided or structured inquiry tutorials into an otherwise standard lecture setting, most ills may be cured. Arguably this holds true if the acquisition of what Duschl has called "final form science" [5] is the sole aim of instruction. However, students learning science should also develop as independent learners. Thus, one may ask, do students learn to learn without strong guidance? Do they learn how to ask relevant questions or analyse more complex problems? Can they synthesise concepts and understand increasingly complex situations?

These practices or competencies cannot be expected to develop as a by-product of instruction designed to serve a different purpose; instead, to achieve this more open inquiry forms of instruction should be employed. Roschelle developed a set of computer simulations that allowed pairs of high school students to co-construct understanding of velocity and acceleration [6]. Typically, they were given a "ball", its trajectory, and a "thin" arrow and a "thick" arrow, to manipulate on a screen; they were not told that the arrows represented velocity or acceleration. The pairs of students were given a number of scenarios that allowed them to pose relevant questions. Many made personal analogies that helped their understanding of the situation. Some discovered that the acceleration arrow changed the velocity arrow, which is a key step in effecting conceptual change from "force as a mover" to "force changes velocity". All of the students made connections between their own ideas and scientific concepts and learned to ask relevant questions and developed ways of answering them [6].

Implications for Pre-service Teacher Education

The various forms of instruction presented that can be used to teach acceleration illustrate a general principle. In a simplistic but useful approximation, didactic teaching methods are best suited to conveying factual knowledge; it helps most students develop operational understanding (in Skemp's words, "rules without reasons" [7]). Structured inquiry approaches may help many students develop conceptual understanding ("knowing what to do and why"). Both forms of instruction align well with students acquiring knowledge and understanding of established facts, theories and approaches. Open inquiry on the other hand aligns with a view of science as improving explanations [8] and valuing students self-constructing knowledge and developing metacognitive skills and agency. Of course, there are no sharp boundaries, and these classifications are to be taken only for illustrative purposes.

In my view, ideally all (secondary school) students should develop factual and conceptual knowledge and the ability to do so unaided; trust their own judgement and make trustworthy judgements; and they should be challenged appropriately in the classroom. Good physics teachers set out to do so in the area of physics, and in doing so, they themselves will continue to develop their own content knowledge for teaching. This tenet underpins my approach to physics teacher education. It is thus paramount for me that PSTs experience, discuss, and understand examples of good didactic teaching, structured or guided inquiry tutorials, and open inquiry approaches.

Despite government policy and national curriculum changes, to date many PSTs have not experienced the full range of instruction forms as a student. The apprenticeship of observation [9] of the majority of our PSTs comprises predominately didactic teaching, or at best pseudo-inquiry. The latter term is not meant pejoratively, but reflects how many teachers have incorporated some ideas of inquiry teaching into their traditional practices of teaching. An example of pseudo-inquiry, often found in Irish schools, is a class that starts with a teacher asking students what they know about the topic on hand; but rather than using this information as something to build the class on, it is used only to prime students and precedes a predominantly didactic remainder of the class. Note that many students and pre-service science teachers alike see nothing wrong with this, and don't see a need for student agency: good teachers know best [10], and good teachers are described mostly in behaviorist metaphors [11].

A Workshop Approach to Physics Teacher Education

Design Parameters

The Irish specification for Junior Cycle Science (for 12-to-15-year-old students) states that "Science in junior cycle aims to develop students' evidence-based understanding of the natural world and their ability to gather and evaluate evidence: to consolidate and deepen their skills of working scientifically; to make them more self-aware as learners and become competent and confident in their ability to use and apply science in their everyday lives" [12]. Given their apprenticeship of observation, it is unlikely that PSTs will be able to support their students' learning unless they develop the mindset, tools and confidence to enact inquiry in their classrooms, when appropriate.

To this end we designed a compulsory physics education module in the second year of a four-year B.Sc. programme at an Irish University that qualifies PSTs as science or mathematics teachers at secondary level. This module focuses on PSTs developing a both a deeper understanding of physics and effective strategies for teaching physics as in an inquiry-based way. The latter raises concerns about unit planning, sequencing units and lessons, choosing a degree of guidance and facilitating collaboration with and between students.

At present the module comprises twelve weekly three-hour slots; in nine or ten of these, we employ a workshop approach to achieve this development. In the remaining slots the PSTs design and microteach a physics class using the techniques they have encountered in the module. Typically, between 10 and 20 PSTs take this module each year. In a typical workshop the PSTs first engage in inquiry activities which are suitable for or easily adaptable to the secondary school setting, and subsequently critique and reflect on the activities [13]. Open-ended activities, almost by definition, can be carried out at any level of education, and can therefore provide an opportunity for our PSTs to acquire physics knowledge at the same time as learning new forms of instruction.

Examples of Workshops

The first four workshops the PSTs encounter in the module are described in detail below. At present, the author is the sole instructor of the module; in the past, the author and a graduate research student would have been instructors.

Workshop 1: Imploding tanker wagon. In the first workshop the PSTs encounter, they are shown a video of a railway tanker wagon that implodes after being steam cleaned. The PSTs are asked to discuss and explain in groups of four why the wagon imploded, and are explicitly asked to sketch the wagon before, during, and after implosion. The instructors limit their interactions with the PSTs to responding to questions of fact (e.g., "Did the vales remain closed? Yes.") and ensuring that the PSTs considered all relevant physics (e.g., "I notice you have been talking about pressure. Are you looking at this at the molecular level?"). A detailed account of the kind of discourse that ensued is given in [14]. In this workshop the PSTs do not critique the activity per se, but rather review a 30 min video recording of how this same activity was taught in a secondary school [15] (the format closely resembles what they have just experienced, but the student discourse that ensues is obviously different). They are specifically asked to note examples of what they consider good teaching, select examples of student ideas they deem interesting and describe how they would respond to these student ideas as teachers.

This approach works well in two ways: it gets the PSTs to critically evaluate their own views of teaching, and it enables us to research these views. In Goodwin's [16] parlance, we get to see what PSTs highlight, and how they code it, to ascertain their professional vision. For example, in the video one student likens the implosion of the wagon to the explosion of an overheated thermometer. The fact that a PST highlights this as an interesting idea reveals something about their views of science teaching; but how they interpret or code it reveals even more. If they write that they would use the statement as a starting point for explaining how the alcohol thermometer expands and evaporates, and fully explain how in the cooling wagon the opposite occurs, a researcher may infer that the PST feels that this class, and possibly every class, should be about students acquiring correct knowledge transmitted by the teacher—in other words, they enact a pseudo-inquiry class. If they state that they would explain the expansion of alcohol inside a thermometer, but leave it up to the students to then construct the explanation for the cooling wagon, a researcher may infer that the PST holds a view that students should construct their own knowledge, but that they feel that this knowledge should be more or less correct at the first attempt, which can be achieved if the steps left for the students ("the conceptual distances") are quite small. If they state that they would encourage the student to come up with a detailed explanation of why the overheating thermometer exploded, a researcher could infer that PST values students developing their own reasoning from bits of knowledge they have already internalized. It goes without saying that the researcher's inferences about PSTs' professional vision should be based on many inputs, not just their written critiques but also their classroom discourse.

It is paramount that the PSTs receive feedback from the instructors in a timely manner that reflects the aims of the workshop. As stated before, our aim is to help our PSTs develop competence and confidence to enact all three forms of instruction illustrated, i.e. didactic, structured and open inquiry forms of teaching. Much of the feedback from the instructors therefore consists of suggestions to PSTs consider other possible options. It may be suggested to the first PST that their chosen (didactic) method of teaching is viable, but that they should think about whether their students will learn to construct knowledge themselves. The third PST may be asked to consider what should happen in the likely case that the student does not arrive at the canonical answer in the allotted class time.

Workshop 2: Classical probability density. As a consequence of our workshop design PSTs have often already learnt something in school or university about the physics topics under consideration. It has been our experience that even PSTs who appear to hold views compatible with or favorable to open inquiry may accommodate these within their existing apprenticeship of observation by suggesting that they would use a preceding class "to introduce all the terms properly". While this is not an indefensible practice, it does go against the ideas supporting and validating the use of inquiry.

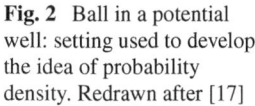

Fig. 2 Ball in a potential well: setting used to develop the idea of probability density. Redrawn after [17]

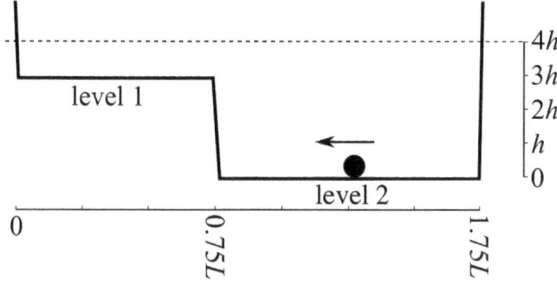

To counter this, we have incorporated a workshop in which the PSTs learn a university level physics topic that they had not encountered before through inquiry. We chose a tutorial developed by Crouse [17] in which students are guided towards the idea of (classical) probability density by considering a ball moving back and forth frictionlessly within an asymmetric two-level system and deriving expressions for the probability per unit length of finding the ball at a particular position (see Fig. 2). The PSTs are expected to use their introductory physics knowledge of kinematics and energy and some basic knowledge of probability. By considering the gravitational potential energy, speed, time spent, and the probability per unit length of finding the ball at each level, students develop the notion of probability density from their own existing knowledge, not from a definition. Within each group PSTs often found different instances when the conceptual step size or their facility with physics concepts and mathematical techniques were problematic, but they typically found that other group members were able to assist them. In this way the tutorial adheres to the tenets of structured inquiry, and provides a good opportunity to discuss different teaching and learning goals including what kind of peer interactions they found useful.

Interesting whole-group discussions may be engendered by assigning different groups of PSTs different learning intentions: some groups may be told the activity is about learning a new topic, others that it is about applying current knowledge in a new setting, others still that it is about experiencing inquiry. This set-up often brings home that there are multiple valid goals for a single activity, and it is worth discussing whether all can or should be pursued at once and to what extent the choice of learning intention would influence the teacher's and the students' behaviors.

Workshops 3 and 4: Forces. As a final example, I will discuss together the next two workshops the PSTs take. The inquiry activities they contain belong together as a single three-week inquiry unit, but they are split into two workshops since there would be insufficient time for the PSTs to critique them otherwise. The worksheets used in the workshop can be found at [18].

The inquiry activity addresses a number of known difficulties with physics. For example, in explaining Hooke's Law, many students (and some teachers) make a statement like "force is mass time acceleration, so the spring force equals mass times acceleration due to gravity". For this reason, the TLS we developed focuses on the concept of balanced forces. As a second example, we know that students often consider that all straight-line graphs should go through the origin and may even equate dividing coordinates to calculating rise-over-run [19], so we require students to engage with linear graphs that do not go through the origin.

The design principles of the TLS are as important as the content. The activities do not take up more time than didactic teaching methods and do encompass all of the related curriculum. The TLS comprises an integrated set of tutorials, practical activities and homework questions that do not require prior introduction of relevant "theory". One tutorial is about representation: it allows students to develop their own representation of forces with arrows. Students extract formulae and discover relationships from their own thinking and their own data—the way most of our knowledge of science has developed. The experiments incorporate a number of questions designed to get students to think about what they are doing. Each tutorial and experiment are accompanied by homework questions aiming to deepen the students' understanding.

Our PSTs' Professional Vision

The PSTs were asked to critique the worksheets they are given at three levels of granularity: in its entirety; episode by episode; and one of those episodes question by question. The global critique allows the PSTs to examine their own views of inquiry; the episode by episode the logical progression; and the question by question critique links the micro-level to the macro-level, and impresses on students that there must be a reason for every question asked.

The PSTs' critiques once again reveal something about their views of teaching science, and omissions may be as revealing as inclusions. A typical global critique may give praise for the entire curriculum being covered, and allowing the students to understand the material so that they will remember it better, without mentioning anything about the intended impact on student learning. Some groups of PSTs, however, at this early stage in their teacher education already think more deeply about other purposes of inquiry, as revealed by this comment:

> We debated on whether introducing this worksheet to either first or second year students. There was some concern that the students may become confused by the questions in the worksheet however [...] In the end we decided that the worksheet would be a good challenge for first year students. The challenge would make them co-operate and be [a] richer learning experience.

Coding individual questions is similarly instructive. Take for example the way two different groups of PSTs coded the following task on the *Forces* worksheet (workshop 3 and 4):

> While the box is moving, a friend comes over to help. He pushes as hard as the first boy. Draw one or more arrows to represent the PUSH force on the box now.

One group's complete commentary on the question,

> The section on introducing more than one force was a little ambiguous in terms of if we were supposed to draw 1 arrow or 2.

suggests that they see teaching as making sure students learn the right information and the right procedures. At the other end of the spectrum, a group noted that

> Actually asking students to design their own diagram also encourages them to think deeply about what it is that they are doing as opposed to interpreting a diagram that someone else has drawn for them.

This group had clearly thought about the purposes of the question in a fundamentally different way. Even this group was not explicit about whether they value the procedure of thinking about representations per se or whether in the end they saw it primarily as a means of better understanding the representation. After four weeks of workshops, the critiques provided by most groups hint at views of science teaching somewhere in between these two positions.

Conclusion

I have described a workshop approach for pre-service science teachers that facilitates three things at once: pre-service science teachers deepen their knowledge of physics, they broaden their views of science teaching, and their instructors gain insight into the development of both. The description of the first four workshops and the comments on some of the critiques exemplifies the approach and suggests that the approach helps most groups of PSTs gain deeper insight into the affordances of inquiry-based approaches to science teaching.

Acknowledgments I wish to thank Leanne Doughty, Paul Grimes, Scott McDonald and Henry Barry for long-standing collaboration on developing and running the workshops, and the many groups of PSTs who have helped the module get better for their successors.

References

1. Harlen, W., Elstgeest, J.: Sourcebook for science in the primary school: a Workshop approach to teacher education. UNESCO, Paris (1992)
2. Shaffer, P.S., McDermott, L.C.: A research-based approach to improving student understanding of the vector nature of kinematical concepts. Am. J. Phys. **73**(10), 921–931 (2005). https://doi.org/10.1119/1.2000976
3. McDermott, L.C.: Oersted medal lecture 2001: physics education research—the key to student learning. Am. J. Phys. **69**(11), 1127–1137 (2001). https://doi.org/10.1119/1.1389280
4. Heron, P.R.L.: Empirical investigations of learning and teaching, Part I: Examining and interpreting student thinking. In: Redish E.F. (eds.) Research on Physics Education: Proceedings of the International School of Physics, pp. 341–350. IOS, Amsterdam (2004)
5. Duschl, R.A.: Restructuring science education. The importance of theories and their development. Teachers' College Press, New York (1990)
6. Roschelle, J.: Learning by collaborating: convergent conceptual change. J. Learn. Sci. **2**(3), 235–276 (1992). https://doi.org/10.1207/s15327809jls0203_1
7. Skemp, R.R.: Relational understanding and instrumental understanding. Math. Teach. **77**, 20–26 (1976)

8. Sandoval, W.A.: Understanding students' practical epistemologies and their influence on learning through inquiry. Sci. Educ. **89**(4), 634–656 (2005). https://doi.org/10.1002/sce.20065
9. Lortie, D.C.: Schoolteacher: A Sociological Study. University of Chicago Press, Chicago (1975)
10. Donnelly, D.F., McGarr, O., O'Reilly, J.: Just be quiet and listen to exactly what he's saying: conceptualising power relations in inquiry-oriented classrooms. Int. J. Sci. Educ. **36**(12), 2029–2054 (2014). https://doi.org/10.1080/09500693.2014.889867
11. Leavy, A.M., McSorley, F.A., Boté, L.A.: An examination of what metaphor construction reveals about the evolution of preservice teachers' beliefs about teaching and learning. Teach. Teach. Educ. **23**(7), 1217–1233 (2007). https://doi.org/10.1016/j.tate.2006.07.016
12. Department of Education and Skills. Junior Cycle Science Specification. Dublin (2015)
13. McDonald, S., Grimes, P., Doughty, L., Finlayson, O., McLoughlin, E., van Kampen, P.: A workshop approach to developing the professional pedagogical vision of Irish secondary preservice science teachers. J. Sci. Teach. Educ. (2019). https://doi.org/10.1080/1046560X.2019.158303
14. Grimes, P., McDonald, S., van Kampen, P.: "We're getting somewhere": Development and implementation of a framework for the analysis of productive science discourse. Sci. Educ. **103**(1), 5–36 (2019). https://doi.org/10.1002/sce.21485
15. https://www.youtube.com/watch?v=Zz95_VvTxZM. Last accessed 2018/12/04
16. Goodwin, C.: Professional Vision. Am. Anthropol. **96**(3), 606–633 (1994). https://doi.org/10.1525/aa.1994.96.3.02a00100
17. Crouse, A.D.: Research on Student Understanding of Quantum Mechanics as a Guide for Improving Instruction. Ph.D. thesis. University of Washington, Seattle (2007)
18. http://castel.ie/junior-certificate-resources/. Last accessed 2018–12-07
19. Wemyss, T., van Kampen, P.: Categorization of first-year university students' interpretations of numerical linear distance-time graphs. Phys. Rev. Spec. Top. Phys. Educ. Res. **9**(1), 010107 (2012). https://doi.org/10.1103/PhysRevSTPER.9.010107

Views and Strategies of Teachers Concerning the Role of Mathematics and Physics in Physics Lessons

Gesche Pospiech⬤, Yaron Lehavi⬤, Esther Bagno and Bat-Sheva Eylon⬤

Abstract Mathematics and physics show a complex interplay. This implies a special role of mathematics in physics as well as in physics education. Here, mathematics seems mostly to be present in the use of formulae and to play an important role in routine problem solving. However, beyond a technical role of mathematics also its structural role, providing insight into the nature of physics, is important for scientific literacy. This aspect leads to the question which views on the interplay teachers have, what guides them in enacting suitable teaching learning processes in the classroom, and which strategies they apply in order to introduce and facilitate the use of mathematics (diagrams and algebraic expressions) in physics lessons. This question was tackled from two sides: inductively by empirical research, and deductively by theoretical analysis. In a first step theoretical considerations led to a model describing specific aspects of pedagogical content knowledge required for adequately teaching the interplay of mathematics and physics. In order to validate it and to identify typical views of teachers half structured interviews with teachers of mathematics and physics of different degree of experience were conducted. The questions were very open in order to explore the field and obtain a wide range of views. From the interviews we could identify several types of teachers' views differing in the importance they give to the technical or structural role of mathematics.

Keywords Physics education · Mathematics in physics · Pedagogical content knowledge

G. Pospiech (✉)
TU Dresden, 01069 Dresden, Germany
e-mail: gesche.pospiech@tu-dresden.de

Y. Lehavi · E. Bagno · B.-S. Eylon
Weizmann Institute of Science, Rehovot, Israel
e-mail: yarlehavi@gmail.com

E. Bagno
e-mail: esther.bagno@weizmann.ac.il

B.-S. Eylon
e-mail: bat-sheva.eylon@weizmann.ac.il

© Springer Nature Switzerland AG 2019
E. McLoughlin and P. van Kampen (eds.), *Concepts, Strategies and Models to Enhance Physics Teaching and Learning*,
https://doi.org/10.1007/978-3-030-18137-6_16

Introduction

Since the times of Galilei mathematics has become more and more central to doing physics. Therefore it is generally undoubted that mathematical elements have to be included in physics teaching in order to achieve the goal that students learn physics and about physics deeply, and that mathematics should play a specific role in physics education on all educational levels. Even students at primary level have their first encounters with mathematical elements such as numbers and units, e.g. meter, kilogram or second. But for many students the use of mathematical elements, especially formulae, seems to cause more fright than fun [1]. Mostly the use of mathematics is connected with the experience of difficult tasks in problem solving and corresponding assessment. Often teachers notice that students are not as successful as expected and detect (surprising) difficulties.

This has led to research on different strategies used by students, e.g. related to problem solving or to the use of representations. These mostly concerned college or university students. A wide range of patterns or strategies, e.g. so-called appropriate and less appropriate strategies, so-called "epistemic games" were uncovered [2–4]. Starting from this finding it seemed useful to look for deeper reasons of the use of inappropriate strategies. One source could be the knowledge and abilities acquired at secondary school or at high school. A study of secondary school students aged 15–16 years old showed that they use very different strategies if faced with non-standard physical-mathematical problems [5]. These strategies are also influenced by the views and attitudes of students at secondary school towards the role of mathematics. It is observed that these views cover a wide range [6].

Changing from the learning perspective to the teaching perspective, the question arises how teachers shape the teaching-learning processes in physics lessons as early as in lower secondary school. As the teachers' views and teaching methods strongly influence the learning processes, understanding and attitudes of students, we are directed to the role of teachers as mediators of the interplay between physics and mathematics. The specific knowledge teachers need for identifying their goals, for shaping their lessons and instructional strategies, and for assessing the learning progress of their students is summarized in the construct of pedagogical content knowledge (PCK).

As the interplay of mathematics and physics covers a special aspect of physics teaching relating two school subjects it seemed appropriate to establish a specific PCK model [7]. In establishing such a model, the distinction between the technical and structural role of mathematics was taken into account [8]. That this distinction might be fruitful was confirmed in a study focused on the epistemic views of teacher students and their relation to the strategies used in problem solving [9]. Their qualitative study showed that a connection can be found in a sense that students who are more aware of the structural role perform better in problem solving. Another study identified groups of mathematics and physics teachers showing that beliefs of teachers are relevant to their choice of teaching strategies [10]. Different characteristic teaching strategies used by experienced teachers could be identified as "Patterns" [11]. We set out to explore which knowledge and which beliefs could reasonably be expected from teachers concerning insight into the interplay and its realization in teaching (for a discussion of the construct of beliefs see Ref. [12]). The benchmark for this would be the views, experiences and knowledge of teachers with many years of teaching experience and with special qualifications, so-called master teachers. First, we will briefly describe the structure of the used PCK model before presenting the interview study.

PCK for Teaching Mathematics in Physics

Teachers operate on the basis of their beliefs shaped by their views on science in general, by their experiences in the classroom, and by their observation of students. Based on these and on the obligatory curriculum they define their goals and objectives. From these they derive guiding principles and choose the appropriate aspects of the interplay of mathematics and physics. These form their strategies and instructional practices by balancing the technical role of mathematics, its structural role, and its role as language. On the whole they develop their own criteria for choosing their focus of teaching. The observation of students leads them to infer the students' views, competences, difficulties and knowledge. These inform the teachers' strategies for introducing and facilitating the use of mathematics in physics lessons, as for instance how they treat mathematical tools, in which way they support students, which role is played by exercising or repeating and how they aim at creating insight into the inter-play of physical structures with mathematical means. The obligatory content or curriculum also plays an important role for the choice of examples and the complexity of content. From this brief description five possible aspects of a model for mathematics-physics specific PCK can be identified (see Fig. 1). A more detailed description of the PCK model and other related models that support it additionally can be found in [7]. This theoretical model forms the basis for an exploratory study.

Research Questions

The main goal of the research presented here is to validate the PCK model described in Fig. 1. Because of the importance of teachers' beliefs for their teaching, one main emphasis was on their perception of the interplay, and how they would try to make it visible for the students. In addition, it is important for teaching success that teachers can diagnose their students' abilities and knowledge. Therefore the focus of analysis was on the following aspects from the model:

- In which way and to what extent do the teachers reflect on the interplay and structural role of mathematics from a general point of view?
- Which are their preferred teaching strategies?

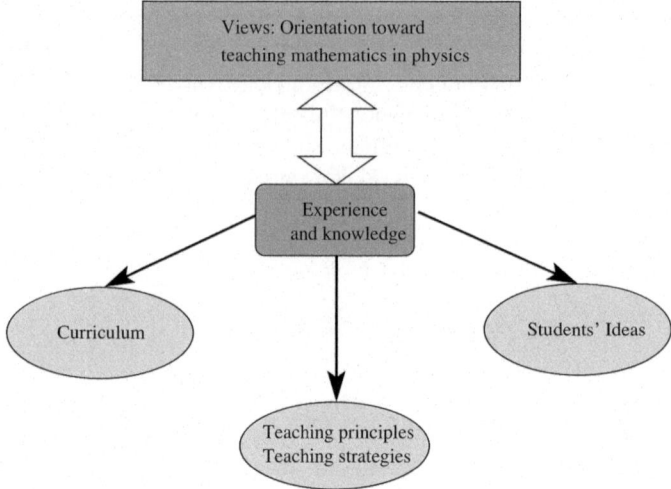

Fig. 1 Basic structure of the PCK model concerning the interplay physics and mathematics

- Are the teachers aware of their students' attitudes, knowledge and competences? How do they describe these?

In answering these questions, we look for differences between teachers with different degrees of experience and qualification. Although we only have a very small sample, we also try to find some underlying characteristics of teachers.

Method, Design and Evaluation

In order to gain a broad insight into the thinking of teachers we conducted semi-structured interviews according to an interview guideline. The related questions were open in order to give the teachers the opportunity to go into depth with their explanations and comments. The questions relevant to this contribution were:

- It is often claimed that the interrelation between math and physics is highly important within the context of physics teaching. How do you regard this claim? Please describe from your point of view which role mathematics plays or should play in physics lessons.
- How do you construct this interrelation within your teaching? What mathematical insights do you use in developing insights in physics? Please provide some examples.
- In your experience, what difficulties do students have in understanding the interrelation between math and physics? Describe, on the basis of your experiences, which problems arise if students apply formulae, diagrams, or explanations.

Thirteen teachers from "gymnasium" were interviewed. Most were known to the researchers in advance and were intended to represent teachers who have the best possible expertise with respect to PCK, especially the interplay between mathematics and physics. All had more than 15 years of teaching experience. Two of them were classified by us as master teachers because of their achievements, the functions they had in the school, or by taking part in developing the physics curriculum of the state. All the teachers were qualified to teach mathematics and physics up to high school level, preparing students for university. But not all of the teachers had taught high school classes in physics recently. Some of the teachers had their main practice in teaching up to grade 10 at "gymnasium".

The teachers got the questions one week in advance in order to be able to prepare themselves. The interviews lasted between 30 and 60 min and were audio-recorded. For analysis they were transcribed completely and were analyzed with qualitative content analysis according to Mayring [13]. We started with deductive categories derived from the PCK model and added some additional categories inductively from the material. In the end we identified 10 main categories and 22 subcategories belonging to 8 main categories, together with 98 codes. About 1100 coded text fragments were coded; some were double coded. Most coded statements concerned teaching strategies, curriculum aspects and perceived problems of students in applying mathematics in physics. This focus might be connected to the explicit inquiry about examples which led to extensive descriptions.

During the analysis the categories and the coding were discussed in the research group. This led to refinement of categories. After one year there was a recoding which only gave minor changes that do not change the overall interpretation of results.

The main categories covered general views of teachers such as "Role of Mathematics in Physics" and the resulting "Teaching Principles". In order to describe in more detail how mathematics could be used from the viewpoint of teachers, we introduced the categories "Technical role", "Structural role" and "Mathematics as Language of Physics". A large category covered "Teaching Strategies". In addition the teachers' perception of students' views was considered in the categories: "Students' Views on Mathematics as Problem", "Students' Views on Mathematics as Supporting" and "Students' Attitudes". In addition, the teachers often referred to the curriculum and gave concrete examples of content and corresponding teaching strategies, so that "Curriculum" formed an additional category.

Results

In order to answer the research questions, we focus on the analysis of the four main categories "*Role of Mathematics in Physics*" (3.1), "*Teaching Principles*" (3.2), "*Teaching Strategies*" (3.3), and "*Students' Views*" (3.4). A few examples of some codings and their place in the PCK model are shown in Table 1.

Table 1 Examples of teachers' statements belonging to some of the main categories

Teachers' quotation	PCK model
That one should teach physics in schools always related to applications	Orientation towards teaching the interplay mathematics in physics
In some places you can arrive only with the help of equations at insights	Teaching principles
Only by this theoretical pursuit, by dealing with these theoretical equations we came to these findings	Orientation towards teaching the interplay mathematics in physics: structural role of mathematics
Now you can see on the basis of this equation (Thomson's formula) what needs to be changed in this resonant circuit	Teaching strategies: showing physical structures with help of mathematics
You first arrive at qualitative statements	Teaching strategies
There are students, for them it is a help to work with formula	Students' views
Students of lower secondary school are afraid of mathematics	Students' views

Teachers' Views on the Role of Mathematics in Physics

All teachers view mathematics as important or very important for physics. However, there were big differences between teachers in the extent of elaboration or depth of reasoning. Some teachers stated in only one sentence that mathematics is important. On the other end of the range one teacher discussed at length six different aspects of the importance. On the whole the view of the teachers is strongly determined by their daily routine and the requirements of the school curriculum. Some teachers seem to concentrate on the technical role of mathematics in stating that it is an "ancillary science". Others stress the importance in the sense of the structural role in saying that at some points (e.g. special relativity) physics understanding can only be reached with help of mathematics. The relation between experiment and theory and the evaluation of experiments with help of mathematics or the predictive power were also addressed. Only one of the master teachers shows a complete picture of the role of mathematics in physics including modeling and indicates that it should be the goal that students at the end of their school career have a physical world view.

Teaching Principles of Teachers

By "Teaching Principles" we mean guidelines the teachers develop on their own, often not explicitly formulated, which they follow in their teaching. These guidelines are influenced by their experience, by the classes they usually teach, the corresponding goals, and their general view on teaching. It would be expected that these underlying teaching principles show themselves in the way teaching is enacted and which role the technical or structural aspects of mathematics in physics play during lessons. The identification of teaching principles were derived from the data, since they had not been expected to arise. It became clear from the data that they serve to bridge the gap between abstract general rules and the concrete planning and actions in the classroom.

We tried to characterize the teachers by their teaching principles. As we have a very small sample this classification has to be interpreted with great care, especially concerning the relative number of teachers in each group. Nevertheless we think that we have captured important types of teachers and their preferences.

The biggest group, comprising about half of the teachers, we called the "application-related teachers". These stress the importance of applications, in order to create a relation of physics to everyday life and practical physics. One teacher stated: "That one should teach physics in schools always related to applications". They see the use of practical examples as an opportunity to motivate students in general, and to apply mathematical means in physics with a goal. They use visualizations or explorations to make physics insightful, independent from the use of formulae.

In the other half of the group we observed different foci. Some teachers can be called "mathematics related teachers". These make a point of carefully explaining the mathematical techniques used in physics, the application of mathematical tools, and their description in physics terms. They base the use of mathematics in physics on the mathematical knowledge of students. The focus is on exactness, e.g. stating "The exactness is important" and on derivations for deepening understanding. Other teachers can be called "concept related". They start more with qualitative descriptions, with experiments or explanations. Teachers in this group try to make sure first that their students understand the concepts of physics before they use mathematical elements or go to problem solving ("I prefer first to bring in an understanding before I treat it with math"). Another group of teachers cannot be sorted into these three. They seem to use very different strategies depending on the circumstances. We can call this group the "multi-faceted" group.

Teaching Strategies

The teachers described sometimes in very great detail how they shape their teaching in using mathematical elements. The most prominent feature was the treatment of proportionality as a core example for dealing with functional dependencies in physics. This is explicitly required in the curriculum and regarded as important. Another central aspect was the use of diagrams, mostly as a bridge between experiment or phenomenon to a quantitative description with help of mathematics. Some teachers, especially in high school, liked to derive additional physical laws from diagrams, e.g. from the area under a curve in kinematics. Teachers with high reflection and regard for the structural role of mathematics stated more often that they use strategies such as "making meaning of formula" or "qualitative reasoning", whereas teachers who tend to teach in a more pronounced math-related way reported more technical-oriented strategies in their teaching examples. On the whole, all the interviewed teachers mainly strive that the students master basic techniques that are important for assessment such as: reading or interpreting diagrams or understanding the path from qualitative statements as expressed in "the more ... the more" to quantitative statements in formulae.

On the basis of teachers' descriptions certain "Patterns" were identified [14]. Such patterns could also be discovered in the descriptions of teachers. The "Broadening Pattern" occurred in basic methods such as using "area under graph" or applying proportionality and its properties in different contexts. The "Exploration Pattern" was sometimes used in exploring the limits of an algebraically expressed law. The "Construction Pattern" used to derive a law on the basis of experiments via the use of different representations most described most often.

In the category "Teaching Strategies" we could also identify several types of teaching. Some teachers focus on the technical aspects as the presupposition for its application in a physical context. Others are seeking a balance between calculating (the technical role) and describing or explaining in order to ensure an understanding of concepts. Of course, some teachers show both types of strategies, e.g. in lower secondary with a focus on describing and in high school more on calculating and problem solving.

Teachers' Description of Students' Views

The teachers indicated observations of students' emotions, achievements and failures. In all these aspects there are many more negative statements than positive ones. There is a clear awareness of an emotional rejection, sometimes even fear, of mathematics in physics by the students. This was explicitly mentioned by seven teachers out of 13. On the other hand, seven teachers also mention positive aspects, e.g. that students prefer tasks with well-defined steps they can simply calculate or that they might be motivated by gaining a new insight through application of a formula. Despite the perceived negative views of the students most teachers describe their efforts to motivate and support students in applying math by the choice of attractive problems, by providing learning aids and systematically repeating important content, by examples from everyday life or by connecting experiments with the mathematical description.

Difference Between Master and Expert Teachers

Our assignment of being a "master teacher" was based on their function in school and school authority. In the sample there were two. When we compared the master teachers with the other experienced teachers it became obvious that the master teachers had more statements on the whole than the others, and above all, a higher fraction of statements from a more general viewpoint, indicating a more reflective view on the role of mathematics. One of the master teachers in particular had significantly more "reflective" statements concerning general goals of physics education, e.g. balancing mathematical and experimental approaches.

Exemplary Description

In this section we describe some of the teachers in more detail in order to indicate more clearly the range of the teachers' views and strategies. We see a certain coherence between the general views and principles and their preferred strategies but without being able to draw a strong correlation.

Teaching with Strong Relation to Mathematics

This teacher has more experience and practice in teaching mathematics than physics. She focuses strongly on the use of mathematics in physics lessons and draws explicitly parallels to mathematics lessons, but shows no deep reflection about the broad picture. She describes in detail her strategies, which often start from experiments or involve them in other ways. A clear connection of physics and mathematics is important to her. She uses different strategies to derive formulae. Besides she is strongly aware of the students' capabilities and problems and takes them into account in her teaching. She has high expectations of the effort and achievement of her students. In order to support them she gives them many examples, opportunities to gain self-confidence in procedures and emphasizes the importance of translating between physics and mathematics.

Teaching with Applications as Main Motivating Factor

Teachers with this focus often use applications or concrete examples as a motivating factor for students. However, clear differences can be seen.

One teacher shows nearly no reflection on the interplay of mathematics and physics. He states that experiments have to be evaluated with mathematical means, supported by graphical calculator. Concerning the attitude of students, he mostly refers to a lack of knowledge or understanding. There seems to be no detailed awareness of their weaknesses or competences.

A different teacher stressed the relevance of applications. He bases his teaching on a theoretical framework by connecting the more general reflections to his shaping of teaching, e.g. by showing the predictive power of mathematics in physics. In his strategies he switches between different approaches: inductive or deductive derivations, using calculations for understanding or making meaning of formula. Additionally he is aware of the attitude, capabilities and problems of his students.

Teaching with Focus on the Interplay

The teacher has a strong perception of the interplay as the interrelation of experiment and theory and about the predictive power of theoretical physics. Mathematics is a tool of physics but also a science in its own right. He prefers to first generate an understanding of concepts before doing calculations. He states that in lower secondary school physics could be taught without calculating, just by describing the relations or qualitative reasoning. He perceives that students tend to start calculating without thinking, have difficulties to interpret diagrams correctly or to do derivations. On the other hand, some techniques are mastered well by the students, e.g. handling a proportionality. He observes that sometimes intuitive understanding of young students gets lost in higher grades.

Teaching with Multi-faceted Strategies

The teacher makes some general remarks about the (technical) importance of mathematics for physics. On the whole no high reflective level can be recognized. Nevertheless he describes different possibilities to arrive at a mathematical formulation of a physical relation, by using experiments or different representations including graphs or mathematical techniques. The diverse strategies serve to adapt to the level of his students. Concerning the students' views, in his observation the cognitive problems are dominant. This teacher is adapting his requirements concerning the interplay of mathematics and physics to the abilities of the classes.

Master Teacher

The master teacher has a high level of theoretical reflection and stresses that teaching, and especially the use of mathematics, should contribute to a physical world view. In his strategies it is important to him that the students relate knowledge of different areas of physics, e.g. by recognizing similar mathematical structures. He weighs his goals and chooses the strategies accordingly and also differentiates his goals according to the age of his students. Concerning the students, he is aware of possible cognitive overload and sees that students tend to focus on calculating and often neglect transfer between physics and mathematics, e.g. they would not compare numerical results with expectations from physics or everyday life.

Conclusions

The analysis of the interviews shows that the PCK model is suited to describe the experience, knowledge and views of teachers that are necessary to teach the role of mathematics in physics. This model and the developed categories could be used to develop an instrument to assess the specific PCK of teachers. It is desirable to be able to identify types of teachers in order to prepare optimized teaching materials and strategies. However, in this study it was not possible to identify a clear relation between the groups formed according to the general view on the role of mathematics, and the groups formed on the basis of the more detailed teaching strategies. In addition, we learned from the interviews that teachers act very strongly on the basis of their experiences with a broad range of views. Therefore, in pre-service teacher education, it seems necessary to foster reflection on the role of mathematics in physics, to show the competences and difficulties of students, and to establish suitable teaching strategies. Therefore the next step would be to develop suitable teaching materials and courses for teacher professional development.

References

1. Angell, C., Guttersrud, O., Henriksen, E.K., Isnes, A.: Physics: frightful, but fun. Pupils' and teachers' views of physics and physics teaching. Sci. Educ. **88**(5), 683–706 (2004). https://doi.org/10.1002/sce.10141
2. Tuminaro, J., Redish, E.: Elements of a cognitive model of physics problem solving: epistemic games. Phys. Rev. Spec. Top. Phys. Educ. Res. **3**(2), 020101 (2007). https://doi.org/10.1103/PhysRevSTPER.3.020101
3. Sherin, B.L.: How students understand physics equations. Cogn. Instr. **19**(4), 479–541 (2001). https://doi.org/10.1207/S1532690XCI1904_3
4. Brahmia, S.M.: Mathematization in introductory physics. Ph.D. Thesis, Rutgers University-Graduate School (2014)
5. Uhden, O.: Verständnisprobleme von Schülerinnen und Schülern beim Verbinden von Physik und MathematikStudent's problems of understanding concerning the translation between physics and mathematics. Zeitschrift für Didaktik der Natürwissenschaften **22**(1), 13–24 (2016). https://doi.org/10.1007/s40573-015-0038-4
6. Pospiech, G., Oese, E.: Wahrnehmung der Mathematisierung im Physikunterricht der Sekundarstufe 1. PhyDid B, Didaktik der Physik, Beiträge zur DPG-Frühjahrstagung (2013)
7. Pospiech, G., Eylon, B.-S., Bagno, E., et al.: The role of mathematics for physics teaching and understanding. Il Nuovo Cimento C **38**(3), 110 (2015). https://doi.org/10.1393/ncc/i2015-15110-6
8. Pietrocola, M.: Mathematics as structural language of physical thought. In: Vicentini, M., Sassi, E. (eds.) Connecting Research in Physics Education with Teacher Education. International Commission on Physics Education (2008)
9. de Ataide, A.R.P., Greca, I.M.: Epistemic views of the relationship between physics and mathematics: its influence on the approach of undergraduate students to problem solving. Sci. Educ. **22**(6), 1405–1421 (2013). https://doi.org/10.1007/s11191-012-9492-2
10. Turşucu, S., Spandaw, J., Flipse, S., de Vries, M.J.: Teachers' beliefs about improving transfer of algebraic skills from mathematics into physics in senior pre-university education. Int. J. Sci. Educ. **39**(5), 587–604 (2017). https://doi.org/10.1080/09500693.2017.1296981

11. Lehavi, Y., Bagno, E., Eylon, B.-S., et al: Towards a PCK of physics and mathematics interplay. In: Fazio, C., Sperandeo Mineo, R.M. (eds.) The GIREP MPTL 2014 Conference Proceedings, pp. 843–853. Università degli Studi di Palermo, Palermo (2015)
12. Pajares, M.F.: Teachers' beliefs and educational research: cleaning up a messy construct. Rev. Educ. Res. **62**(3), 307–332 (1992). https://doi.org/10.2307/1170741
13. Mayring, P.: Qualitative Inhaltsanalyse. In: Handbuch qualitative Forschung in der Psychologie, pp. 601–613. VS Verlag für Sozialwissenschaften (2010)
14. Lehavi, Y., Bagno, E., Eylon, B.-S., et al: Classroom evidence of teachers' PCK of the interplay of physics and mathematics. In: Key Competences in Physics Teaching and Learning, pp. 95–104. Springer (2017)

Part IV
Bridging Gaps in Student Motivation and Engagement in Physics

Project Accelerate: Closing the Access Gap to Physical Science Careers and Academic Programs

Mark D. Greenman ⓘ

Abstract Economically disadvantaged and underrepresented high school students in many urban, rural, and small suburban communities don't have access to Advanced Placement® (AP®) courses. Lacking opportunity to access rigorous physics courses in high school, these demographic groups are hard pressed to compete in physical science related STEM fields and academic programs with their peers from more affluent communities. Project Accelerate, a National Science Foundation funded project, is a partnership program between Boston University (BU) and high schools to bring a College Board accredited AP® Physics 1 course to schools not offering this opportunity. Preliminary results indicate that students participating in Project Accelerate do as well or better than their peers enrolled in traditional classroom based AP® Physics 1 classes.

Keywords Digital learning · Advance placement · University partnership · Physics education research

Introduction

Economically disadvantaged and underrepresented high school students in many urban, rural, and small suburban communities don't have access to Advanced Placement® (AP®) courses either because of a lack of trained teachers, limited or no AP® program, or a school history of low participation. Physics is often a "gate keeper" course to entry into Physical Science, Technology, Engineering and Mathematics (STEM) careers and academic programs. Lacking opportunity to access rigorous physics courses in high school these demographic groups are hard pressed to compete in STEM fields and academic programs with their peers from more affluent communities. Project Accelerate is a partnership program between Boston University (BU) and the high schools combining the supportive infrastructures from the students'

M. D. Greenman (✉)
Physics Department, Boston University, Boston, MA 02215, USA
e-mail: greenman@bu.edu

© Springer Nature Switzerland AG 2019
E. McLoughlin and P. van Kampen (eds.), *Concepts, Strategies and Models to Enhance Physics Teaching and Learning*,
https://doi.org/10.1007/978-3-030-18137-6_17

195

traditional school with a highly interactive private edX online instructional tool to bring a College Board accredited AP® Physics 1 course to schools not offering this opportunity. During the 2015–2016 academic year, Boston University piloted this model with four Boston Public School (BPS) high schools and three small suburban high schools. During the first year of the pilot, students enrolled in Project Accelerate outperformed their peer groups enrolled in traditional AP® Physics 1 classrooms.

The Problem

There is a critical need to develop STEM competencies among youth from demographic groups underrepresented in the STEM workforce. While underrepresented youth make up more than 50% of today's American high school population, African–American/Black and Hispanic/Latino youth each comprise only 7% of STEM graduates and 6–7% of the STEM workforce [1]. Underserved high school graduates are just as likely as non-underserved populations to be interested in STEM—49% in each case. However, underserved students are far less prepared for college STEM coursework than are students overall (e.g., only 25% of underserved STEM students met the ACT College Readiness Benchmark in science compared to 59% of students who are not underserved). These data indicate that a program to increase academic readiness can succeed in increasing participation in STEM baccalaureate and career pathways [2].

Evidence exists that students who score 3 or higher on AP® exams have greater success in college than students who did not take an AP® course. However, students who receive scores lower than 3 do not perform noticeably better than a comparison group of high school students who did not take a STEM AP® course [3–5]. This indicates how critically important quality curriculum, prepared teachers, and appropriate scaffolding are to student success [6].

The most recent reports indicate schools with predominately low-income students, both rural and urban, lag in AP® offerings by a 2:1 margin and underrepresented groups lag in taking these courses even when offered by a factor of 2 compared to whites and 4 compared to Asians [7, 9]. There is also recent evidence that in schools that do offer AP® programs, there is a large gap in participation between low- and high-income students, regardless of race. What is clear is that economically disadvantaged and underrepresented student groups share an equal interest in STEM as non-underserved students, but they are too often lacking the opportunity to access these gateway courses to success in physical science college programs and STEM careers [2]. Robinson et al. [8] has shown that "taking advanced courses in mathematics and the sciences in high school, e.g., AP® courses, is good preparation for university work in engineering and other STEM careers". More recently, the State of California and the College Board have collaborated on bringing more AP® courses to underserved students. The latest data indicate that a large fraction of underrepresented students (30% or 8800) could potentially succeed in AP® STEM courses but are not enrolling due to lack of opportunity [9].

Boston Public Schools (BPS) in Massachusetts USA is a prime example of this problem: Locally, the Boston Public School system, a typical American urban school system, has 34 high schools serving a district student population of 51,000 students. Of these 34 schools, based on the Massachusetts Department of Education WEB site, only 4 high schools during the 2015–16 academic year offered algebra based AP® Physics 1, the curriculum supported by Project Accelerate. A total of only 151 BPS students took the AP® Physics 1 exam during the 2015–16 school year. Of the 151 students who took the AP® Physics 1 exam, in a traditional classroom environment, only 8% earned a 3 or better. The BPS AP® Physics 1 passing rate is less than one-quarter the Massachusetts state average of 43%. Boston Public Schools, with demographics of 75% black or Hispanic and nearly 100% on free or reduced lunch programs shows an AP® Physics 1 profile score similar to the National AP® Physics 1 scores for under-represented minorities. The success rate (score of 3 or higher) nationally for black and or Hispanic students taking the AP® Physics 1 exam during the 2015–16 academic year was 16%. AP scores are reported on a 5-point scale with scores of 3, 4 and 5 defined by the College Board as qualified, well qualified and extremely well qualified [10].

Project Accelerate

Project Accelerate is a *partnership* between Boston University (BU) and high schools providing a structured, supportive, and rich educational opportunity for underserved students. Project Accelerate is a potential scalable and sustainable solution to closing this access gap to STEM careers and academic programs.

Four Components to Project Accelerate: Project Accelerate combines four components to support student success: (i) An interactive edX Small Private Online Tool; (ii) The supportive infrastructures of the partner high school; (iii) The coordination and academic support of the university partner; and (iv) A hands-on laboratory option.

Online Instruction Tool: The online instructional tool is supported through the edX platform. EdX was founded by Harvard University and MIT in 2012 as an online learning destination. Today there are more than 90 global partners, including Boston University, in the edX online provider community. The edX online instructional tool is short on "video professor segments". Instead, students are engaged throughout the online instructional tool with interactive explorations using Direct Measurement Videos by Peter Bohacek, PhETs from University of Colorado and interactive HTML5 simulations by Andrew Duffy co-PI on Project Accelerate. Videos when included are no longer than 7 min and are provided as an alternative learning modality reviewing instruction provided through engaging simulations, Direct Measurement videos and interactive instruction. The online component is authored specifically to work seamlessly with a typical high school schedule. There are 28 graded virtual laboratories, 28 graded homework assignments, 24 graded quizzes and 8 graded, proctored and timed simulated AP® style tests. The end of term tests are proctored by the partner building liaisons and all assignments are graded through the edX online instructional tool. Participating students pose queries and engage in discourse with Project Accelerate staff and the larger student learning community through an online discussion forum.

High School Partner: The HS appoints a professional staff member to serve as "HS liaison" (e.g., science teacher, outreach coordinator, or guidance counselor). The HS liaison facilitates communication between the school, students and the project team. The HS is provided a set of guidelines for enrolling students (i.e., maximum of 10, algebra 2 proficient, potential for independent learning, demonstrated history of submitting assignments in a timely fashion and interest in academic challenge), but is provided a good deal of latitude in vetting students into the program. The high school assigns participating students in-school time like any other major course and includes the course on the student's transcript and report card. The HS liaison is the chief encourager and nagger keeping students on task and on schedule. The HS liaison does not provide formal content instruction.

University Partner: The Project team appoints a "University liaison" who coordinates all aspects of the program. The University liaison monitors student performance, and communicates regularly with the HS liaison concerning issues that might impact student performance. The University liaison provides formal midterm reports, end of term grades and end of course grades. The University liaison monitors discussions on the online forum and where applicable supervises undergraduate Teaching Assistants who facilitate the on-campus hands-on laboratory component of the course.

Hands-on Laboratory Option: Students within commuting distance to the university are required to attend weekly small group 2½ h laboratory sessions on the university campus. Sessions give students an opportunity to explore concepts through hands-on inquiry based laboratories, receive additional support based on individual learning needs and exposure to a university campus. Sessions are facilitated by trained and supervised undergraduate Teaching Assistants—physics undergraduates trained in STEM pedagogy and physics pre-conceptions through a 2-credit, one-semester course. School partners not within commuting distance are encouraged and wherever possible supported in providing students with a hands-on laboratory component to complement the online instruction tools. Partner schools, including our commuting partner schools, offering a significant and quality hands-on laboratory component report the course as an accredited College Board AP® Physics 1 course. Other partner schools record the course as AP Physics 1 Preparation. However, all students are required to register and complete the AP® Physics 1 exam.

Research Agenda

Our research agenda explores three aspects of the program; (i) the efficacy of the program, (ii) the scalability of the model, and (iii) the long-term sustainability of the program.

Efficacy: We explore program efficacy by measuring student outcomes through AP® exam performance, pre/post scores on the Force Motion Concept Evaluation (FMCE), course completion rate, impact on student STEM choices and longitudinal college and career choices and performance.

Scalability: We additionally explore the structure of the blended model and university to school partnership both in the local pilot program, and in terms of whether it can be replicated at other sites around the country.

Sustainability: We look at long-term sustainability through a cost analysis of delivering the program at a price point that would likely be attractive to school administrations.

Pilot Project

Student Demographics: During the 2015–2016 academic year, BU partnered with 7 high schools in 4 districts to bring a blended AP® Physics 1 course to underserved secondary school students who would otherwise not have access to AP® Physics. Our partner schools included 4 Boston Public School (BPS) high schools and 3 small suburban high schools. None of the participating schools offered their students the opportunity to enroll in AP® Physics. A total of 24 students enrolled in this pilot project. Seventeen attended 4 high schools in the BPS system, and 7 attended 3 schools in central and western Massachusetts. The demographics for our first cohort were 67% black and Hispanic and 75% on free and/or reduced lunch programs.

AP Physics 1 Exam Results: All participating students were required to take the College Board AP® Physics 1 exam. Although the sample size was too small to provide statistical significance, preliminary results are very promising. Project Accelerate students did as well or outperformed their peer groups enrolled in traditional AP® Physics 1 classrooms. Fourteen percent of the BPS students completing the Project Accelerate program scored a 3 or better compared to 8% for BPS students enrolled in traditional AP® Physics 1 classrooms. 71% of the non-BPS students completing the Project Accelerate program scored a 3 or better compared to 43% for non-BPS Massachusetts students enrolled in traditional AP® Physics 1 classrooms.

FMCE: We administered the Force Motion Concept Evaluation (FMCE) as a pre/posttest. This instrument is used by many universities and colleges to gauge the learning gains of students within their own introductory college level physics courses. Students in Project Accelerate had a paired fractional gain of 0.53 which is considered very good by the physics education research community.

Overall Retention: Twenty-one of the initial 24 students enrolled in this program completed the course—resulting in an 88% retention rate. At week 7, a student withdrew and commented, "The course is more work than I want to do. I am a senior". A second student withdrew in week 15 with the comment, "Just not comfortable having to direct my own learning. I prefer having the teacher tell me what to do while I'm in class". A third student, dealing with personal issues, withdrew in week 17.

Student Attendance: BPS students attended a weekly 2½ h laboratory block held from 4 to 6:30 p.m. on the BU campus. The attendance rate for BPS students at these laboratory sessions for the full year was 90%.

Student STEM Interest: Eight of our 14 Boston Public School completers applied to participate in summer STEM programs. Of these 8, 6 indicated in our post-course survey that participation in Project Accelerate was either "very important" or "somewhat important" in their decision to apply to a summer STEM program. The remaining 2 of those applying to participate in a summer STEM program indicated they were planning on applying to such a program prior to entering our course. 52% of all students indicated on the post-course survey that they were either "much more likely" or "somewhat more likely" to pursue a STEM program in college as a result of their participation in Project Accelerate. The remaining 48% of students indicated "no impact on their decision," and no student chose the two negative choices of "somewhat less likely" or "much less likely" to pursue a STEM program in college.

Scaling up and Replication

More Partner Schools and a Replication Site: Project Accelerate is a National Science Foundation (NSF DUE 1720914) funded project. With NSF support over the next three years, we will be offering Project Accelerate to an expanded number of partner schools and supporting several replication sites. All schools partnering with Project Accelerate during our pilot year have requested to be part of the program again—a vote of confidence in the program. During the coming academic year, Project Accelerate will more than double in size, partnering with a total of 16 high schools, enrolling 67 students and opening our first replication site. The project will include 6 BPS public and public charter schools, 4 other Massachusetts schools, 5 West Virginia high schools and 1 New York City high school. The 5 West Virginia schools will be supported through our first replication site partner West Virginia University.

Conclusion

Project Accelerate offers a potential solution to a significant national and international problem of too few underserved young people having access to high quality physics education, resulting in these students being ill prepared to enter STEM careers and STEM programs in college.

Project Accelerate is based upon the compelling need to provide access to AP® Physics for economically disadvantaged and other underserved groups. Research shows that providing high quality education is critical to student's success in the future, and on the growing body of evidence that blended course structures, combining online learning with in-person sessions, can be very effective in improving student learning outcomes (see a review by Means, [11]). In addition, several studies have demonstrated that technology improves access to information, and hybrid or blended models engage students more effectively [12–15].

Thousands of high schools do not provide opportunities for underserved students to access AP® physics [9]. Project Accelerate blends together the supportive formal structures of the student's home school, immediate acceptance into school curricula through the AP® designation, a private online instructional tool designed specifically with the needs of underserved populations in mind, and small group laboratory experiences to make AP® Physics accessible to underserved students.

Finally, Project Accelerate is a scalable model of STEM success, replicable at sites across the country, and therefore setting up for success thousands of motivated but underserved students every year.

Acknowledgements The pilot program was funded by the Digital Learning Initiative program at Boston University. Project Accelerate is currently funded by the National Science Foundation Division of Undergraduate Education (NSF DRL #1720914). Any opinions, findings, and conclusions or recommendations expressed in this material are those of the author(s) and do not necessarily reflect the views of the National Science Foundation.

References

1. Landivar, L.C.: Disparities in STEM employment by sex, race, and hispanic origin. National Census Report (2013). https://www.census.gov/library/publications/2013/acs/acs-24.html. Last accessed 30 Sept 2017
2. ACT: Understanding the Underserved Leaner (2014). https://www.act.org/stemcondition/14/pdf/STEM-Underserved-Learner.pdf. Last accessed 30 Sept 2017
3. Scott, T.P., Homer, T., Yi-Hsuan, L.: Assessment of advanced placement participation and university academic success in the first semester: controlling for selected high school academic abilities. J. Coll. Admiss. **208**, 26–30 (2010)
4. Klopfenstein, K., Thomas, M.K.: The advanced placement performance advantage: fact or fiction. Texas Christian University and Mississippi State University, American Economic Association Annual Meeting Papers (2005)
5. Sadler, P.M., Tai. R.H.: Advanced placement exam scores as a predictor of performance in introductory college biology, chemistry and physics courses. Sci. Educ. **16**(2), 1–19 (2007)
6. Klopfenstein, K.: Recommendations for maintaining the quality of advanced placement programs. Am. Second. Educ. 39–48 (2003)
7. Handwerk, P., Tognatta, N., Coley, R., Gitomer, D.H.: Educational Testing Service. https://www.ets.org/Media/Research/pdf/PIC-ACCESS.pdf (2008). Last accessed 30 Sept 2017
8. Robinson, M., et al.: AP mathematics and science courses as a gateway to careers in engineering. In: Frontiers in Education IEEE (2003)
9. College Board: https://www.collegeboard.org/california-partnership/ap-expansion (2015). Last accessed 30 Sept 2017
10. College Board: AP Web Site. What is an AP Test Score and What Does it Mean? https://apscore.collegeboard.org/scores/about-ap-scores/. Last accessed 30 Sept 2017
11. Means, B., Toyama, Y., Murphy, R., Bakia, M., Jones, K.: Evaluation of Evidence-Based Practices in Online Learning: a Meta-Analysis and Review of Online Learning Studies. U.S. Department of Education. www2.ed.gov/rschstat/eval/tech/evidence-based-practices/finalreport.pdf (2010). Last accessed 30 Sept 2017

12. Delialioglu, O., Yildirim, Z.: Design and development of a technology enhanced hybrid instruction based on MOLTA model: Its effectiveness in comparison to traditional instruction. Comput. Educ. **51**(1), 474–483 (2008). https://doi.org/10.1016/j.compedu.2007.06.006
13. Sanders, D.W., Morrison-Shetlar, A.I.: Student attitudes toward web-enhanced instruction in an introductory biology course. J. Res. Comput. Educ. **33**(3), 251–261 (2014). https://doi.org/10.1080/08886504.2001.10782313
14. Gunter, G.A.: Making a difference: using emerging technologies and teaching strategies to restructure an undergraduate technology course for pre-service teachers. Educ. Media Int. **38**(1), 13–20 (2001)
15. Tuckman, B.W.: Evaluating ADAPT: a hybrid instructional model combining web-based and classroom concepts. Comput. Educ. **3**, 261–269 (2002)

"There Are no Things Inside Things": An Augmented Lecture to Bridge the Gap Between Formal and Informal Physics Education

Marco Giliberti◉

Abstract After 13 years of research about how to make scientific theater a power-ful tool in physics education (see for instance: Carpineti et al. in Nuovo Cimento B 121:901–911, [1]; Carpineti et al. in JCOM 10(1):1–10, [2]; Barbieri et al. in Il Nuovo Cimento, 38C:1–10, [3]), the group "Lo Spettacolo della Fisica" (The Physics Show) (http://spettacolo.fisica.unimi.it) of the University of Milan, together with Stefano Oss of the University of Trento and Andrea Brunello of the Arditodesio Theater Company (Trento, Italy) created and developed the idea of "Augmented Lectures" (ALs). These are lectures performed in a theatrical setting with the help of profes-sional actors (and/or of painters, musicians, cooks, etc.) with the aim not only to create fascination and passion for physics, but to make people think and reflect on some specific disciplinary topic as well. Although ALs do not need any specific dis-ciplinary preparation, they can, nonetheless, be seen as complementary lectures that can be part of a secondary, or even of a university, formal path in physics. In the aug-mented lecture titled, "There Are No Things Inside Things" (TANTIT), the theme "what is understanding?" is tackled starting from the proprieties of a magnet and developed with the analysis of the double slit experiment in different settings: from the "classical" photons and electrons experiments, up to those with atoms, fullerenes and tetraphenylporphyrin. Experiments with neutrons and interaction-free measure-ments are also analysed. At the basis of quantum mechanics, there is elegance and simplicity; and it is precisely the simplicity of this theory—clearly related to its pow-er—that upsets us, because, on the contrary, the world we are used to is complicated. Some of the peculiarities of quantum physics, and of the conceptual challenges they present to the descendants of fruit pickers and animal hunters as we are, are there-fore a stimulus to discuss the meaning of explaining and understanding, in physical sciences (Cavallini, Giliberti in Epistemologia 31(2): 219–240, [4]). A physicist (the author) and an actor (Giacomo Anderle) act together with real data, experi-ments, songs, dances, equations—also involving the audience on stage—having in mind that culture is culture only if it changes your life. The aims and the struc-ture of the augmented lecture TANTIT firstly performed inside the "Teatro della

M. Giliberti (✉)
Dipartimento di Fisica, Università degli Studi di Milano, Milan, Italy
e-mail: marco.giliberti@unimi.it

© Springer Nature Switzerland AG 2019
E. McLoughlin and P. van Kampen (eds.), *Concepts, Strategies and Models to Enhance Physics Teaching and Learning*,
https://doi.org/10.1007/978-3-030-18137-6_18

Meraviglia" (Theater of Wonder) festival in Trento (www.teatrodellameraviglia.it/)
will be presented and briefly discussed together with its potentialities.

Keywords Theater · Quantum physics · Formal and informal education

Introduction

The Physics Show Project (TPS) was born in 2004 from an idea of Marina Carpineti,
Marco Giliberti and Nicola Ludwig, three researchers of the Physics Department of
the University of Milan. The project started in Italy when the number of students
enrolled in scientific degree programs reached a minimum. TPS had (and still has)
the desire to promote physics, with the aim to be a remedy for the general perception
of physics not only as a difficult—as, in fact, we must admit it is—but also as an
alien subject, far from everyday life, if not for its technological aspects, and, above
all, very far away from personal needs.

The activity of TPS is well documented by seven shows (see Figs. 1, 2, 3 and 4),
the two lecture-shows and the three ALs, all written and performed by the creators of
TPS (sometimes with other co-authors), that to date (4th September 2018) reached
an audience of about 135,000 people in about 400 performances. The TSP project
has also been analysed in various scientific publications [see for instance: 1, 2, 5–10].

The purposes of these shows are, mainly, to innovatively promote a physics culture,
to answer a demand for renewal of physics education, and to develop activities that
brings out the appeal, the creativity of physics and that leads to surpassing the standard
textbook presentation of physics.

Fig. 1 One of the TPS shows: "Alice nel paese dell'energia" (Alice in energy-land); A poster for
the "Piccolo Teatro di Milano—Teatro d'Europa" (left); a picture from the show (right), from left
to right: Marco Giliberti, Nicola Ludwig and Marina Carpineti. *Credits* Marco Rossi

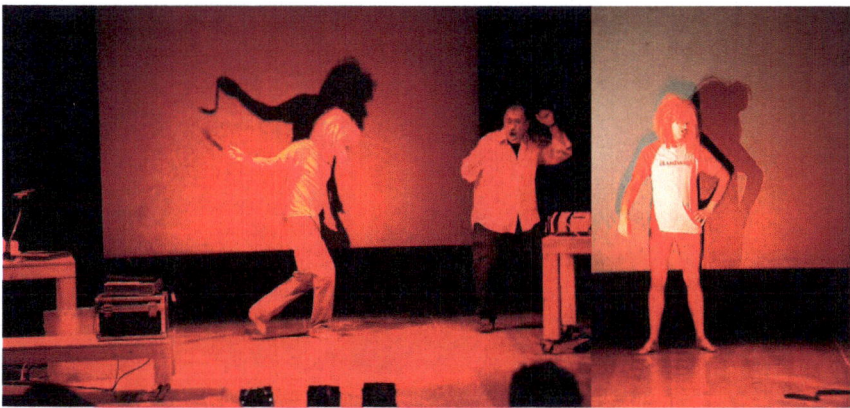

Fig. 2 The TSP show "Luce dalle stelle" (Light from the stars) performed at Teatro Franco Parenti, Milano; striptease of Nicola Ludwig to show the vision under monochromatic light. *Credits* Fabrizio Favale

Fig. 3 Pictures taken from the TSP show "Alice nel paese della scienza" (Alice in science-land); performing: Marina Carpineti, Marco Giliberti and Nicola Ludwig. *Credits* Fabrizio Favale

In particular, among the fixed points that, as guidelines, structure TPS' shows, we want to point out the general decisions to avoid:

- disciplinary explanations—to not turn shows into lectures;
- playing out biographies of scientists—to stronger and properly stage science as the leading character of the show;
- popularization—that, in the best-case scenario, conveys notions by simplifying concepts in an everyday, often inappropriate, common language.

Since in this paper we deserve special attention to ALs, their purposes will be discussed separately in the next sessions.

Fig. 4 "Facciamo luce sulla materia" (Let's shed light on matter) by TPS, performed (from left to right by Marina Carpineti, Nicola Ludwig and Marco Giliberti) at Il Piccolo Teatro Studio (Milano). *Credits* Marco Rossi

TANTIT

The importance of science education both in and out of schools, as well as in higher education, has been much debated worldwide. While the desire to understand is completely natural, the ability/possibility to understand is not, in general, spontaneous; it must be learned and internalized. In order for knowledge and culture to become a personal property, individuals must be able to internalize it in a personal way. Yes, but how? School cannot be enough for this. The culture of a society cannot, in fact, be compared only with the school culture. And indeed, a school that is increasingly open to society needs a society to be equally open to school. The environment in which we live is fundamentally important for the formation of the person and, therefore, in order to insert "correct" physics into a broader culture, it is important to provide a "correct" social image of physics too. In other words, to obtain deeper learning of physics in a formal context (i.e. at school) it is very important what is spoken, read, seen and experienced in contexts which *are not* formal. This implies that the distance between formal, non-formal and informal education has to be smoothed and a suitable combination of these three "forms" of education has to be reached. Theater can be very useful for this purpose; it can develop scientific fantasy and help improve learning using emotions more effectively, thus developing the physics knowledge through the channel of affectivity, enhancing the needs of personal culture and eliminating, therefore, cultural discrimination and gender; moreover, promoting a culture that is more deeply scientific and, precisely for this reason, even more human. On this front, the European Commission has recently recognized theater—generally seen as a way to implement only informal or at least non-formal education—and showmanship as a way for approaching Inquiry based Science Education (IBSE) and implementing it in classroom. For example, through the European project "TEMI" (Teaching Enquiry with Mysteries Incorporated) [11–13] that involved the Physics Department of the University of Milan together with 12 other partners. However, despite the European Commission efforts, the bidirectional links between informal and formal science education need further research to better understand their nature and effects and influence the way these can be implemented both in school and society.

In line with bridging formal and informal learning, a different kind of approach to physics has been proposed by TPS and has given rise to the development of ALs. These are real lectures on a specific disciplinary topic that are performed in a theatrical setting with the purpose of making people think, reflect and discuss physics; this time with prompts coming from *outside the school*. Nonetheless, ALs are not intended to popularize concepts in common language, but rather to make use of the full range of "technical devices", that is: experiments, images, formulae and graphs; so as to give a multimodal flavour of a "deep" physical insights. In fact, as Sergio Escobar (director of the "Il Piccolo Teatro di Milano - Teatro d'Europa", and graduate of philosophy of science) said, "*Translating by popularizing is tantamount of betraying both, the science and the common man. Therefore, it is necessary to create a new relationship of confidence that is not reassuring but 'intriguing' between the science and the common sense. Theater can do very much for that*" [14].

In the following section, we will give a brief account of the structure of the script and the aims of the AL entitled "TANTIT".

Structure of the Script

The TANTIT lecture/show is a dialogue between Marco (the author of this paper) who plays the role of a physics professor, and Giacomo (a professional actor) that plays the role of a physics student who finds it difficult to pass examinations. They meet by chance in a classroom to start preparing a low-cost welcome party for the new Rector of the University. In a short time, they start discussing physics experiments using real data, and let themselves be led by dances, songs, poems, and pieces coming from the literature; fun gags come out, with the direct participation of the theatre audience.

As stated previously, this 70 min performance does not require any previous knowledge of physics and it is, therefore, suitable for a general audience. Until now, "TANTIT" has been performed inside the "Teatro della Meraviglia" festival in Trento and at the *Aula Magna* of the Faculty of Engineering in Palermo where it gave birth to a study about its effect on the public and in particular in relation to the views of the Nature of Science (NoS) of the audience [15]. The show is designed as a means of reflection on mankind, on the nature of scientific language and on the "physics explanation" of the universe (themes that are all particularly suited for high school students and beyond).

A brief digest of the script, together with some excerpts, is presented so as to elucidate the topics encountered and the way they are discussed.

Scene 1

Giacomo (G) and Marco (M) enter the scene together, talking to each other. Marco gives Giacomo a sack of potatoes to be peeled for the party.

G: *speaking about the difficulties he founds in exams.* In those moments, it often happened—often … almost always, to be precise—to be standing there; straight as a pole, having to say something. But every time, at that precise moment, the jaw snapped, the mouth tightened and then I misted up and was silent. Mute as a fish. Students looked at me, waiting for a clue of thought, sympathized for me at first, but then "this here is stupid" was heard in a low voice, and students looked somewhere else.

[…] I did not bat an eyelash, I did not sweat, I did not turn pale, but inside I was a tumult, inside I was all the unexpressed words, the things I would say, and also those I would have kept silent.

[…] "What do you mean?" The professor asked. "Tell me it in simple words". Simple words… Simple words… Everything is simple. But, since I did not even know what was happening to me, what was there, in that separate and inaccessible place, surrounded by a wall, how could I explain it? Maybe sometimes there are no simple words. Maybe sometimes there are no words at all. And so? How to understand?

M: Understanding, what does it mean? And explaining? What does it mean? Very often, explaining something is making an analogy with what we already know, with things we are familiar with. However, about familiar things, we almost never wonder why, it seems that they do not even need to be explained.

[…] That a box with six bottles of water weighs more than a box with only one does not need to be explained. Explaining means doing an analogy with things we are used to and that, hence, we usually find simpler.

[…] Physics puts into crisis this way of proceeding. In particular, quantum mechanics dramatically highlights the limits of this conception of our interpretation of the world.

[…]

Showing a magnet and playing with it. This is a magnet. It attracts pieces of iron or another magnet. At its extremities, it has two poles, north and south, separated. Even if we break the magnet into two halves, [*he breaks the magnet into two pieces*], we cannot divide the two poles, rather we get two more magnets, each one with a north and a south. It does not happen, as for these building blocks, that we can separate the reds from the blue [*playing with the blocks*]. In the case of the magnet, each piece is still "red" on the one side and "blue" on the other. Not only: each piece has about the same "power" as the original magnet.

Interesting.

If we think about its structure, we will immediately understand that we can consider it as consisting of many small magnets, each with a north and a south. In fact, in this way, we can explain the observed properties very well, for example that if we break the magnet, we do not separate the poles.

But, is this an explanation?

[…]

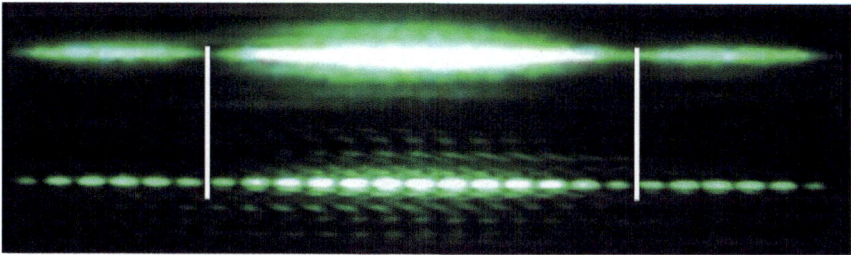

Fig. 5 A slide comparing high intensity single and double slit experiments made, especially for "TANTIT", with a green laser

If we think about it for a moment, we will immediately realize that explaining something complicated has the meaning of modelling it in terms of things that have to be simpler, that is, things which have to have less properties than those we want to explain …

The point is that not always—to tell the truth almost never—the things we are used to are, in this sense, simpler.

Scene 2

The dialogue continues with a discussion on the classical and the quantum version of the double slit interference, with slides and videos of real experiments performed with light and electrons displayed (see Fig. 5).

Scene 3

The importance of mathematics in modelling physical situations is presented, starting from Maxwell's equations, by means of a song (the author of which is the same Maxwell) that is sang by Giacomo and by a dance (see Fig. 6, left).

Interferometry with atoms, fullerenes and tetraphenylporphyrin is presented too, while "which-way experiments with neutrons" and interaction-free measurements are analysed, also involving on the stage people from the audience (see Fig. 6, right).

Scene 4

After some difficulties that arose in understanding single-quantum interferometry experiments…

G.: Ah! Ah! Ah! They make fun of us, a joke, they are joking! Ah! Ah! Ah! I cannot believe it!
I cannot believe it? But yes! Now I understand! What you are telling me is that what we see, touch, and feel is not reality. Or rather, it is not the whole reality; there are other realities that we cannot see, nor even imagine. And we cannot imagine them because the words and the images that we have at our fingertips bring us back to the known, to the concrete, to the senses. They bring us down to the earth as the force of gravity […].

Fig. 6 A dance (made, from left to right, by Giacomo Anderle and Marco Giliberti) to present Maxwell's equations in "TANTIT" (left); public on stage to discuss interaction-free measurements (right). *Credits* Stefano Oss

M.: Our mind struggles to imagine what I am telling you because our understanding passes through the language; that gives names ... of things when there are constant properties; that expresses the changes through verbs, and the accidents with adjectives ... And that is why just one word, like photon, or atom, makes the image, of something like a ball or a of small planetary system, materialize in our mind. Complicate things ...

Almost always the language describes images. Therefore, it is clear, as Wittgenstein writes, that if we want to understand the meaning of what we say, we must explore the images. But images already allude to a particular job, they are stronger than reasoning, and so they make fun of us.

This does not mean that what we are describing is complicate, it is only far from what we are used to. I would not like to give you the idea that we do not understand these phenomena. Rather the contrary! The theory that describes them, Quantum Electrodynamics, is the most precise theory that humanity has ever formulated! But it is difficult to express it in words, because words describe images, which make fun of us [...].

In discussing the necessity of abstract concepts ...

Our life is full of abstract words that do not refer to things, and make it beautiful, interesting, personal. Like: love, brotherhood, friendship, beauty, loyalty. Concepts without which it would be difficult to express ourselves, even difficult to live [...].

Scene 5

After other dialogues between Marco and Giacomo, that we do not report here so as not to spoil the surprise of seeing the AL, the lesson ends with the following words of Marco.

M.: Giacomo, in a way that is similar to the things of the world around us, which we really do not understand, but we are used to them, we can say that even in physics, we do not basically understand, but get used to it!

And anyway, as the Japanese physicist Masaki Hori says: "Sometimes it's more important to peel potatoes than think about the universe"; so let's start peeling potatoes.

The AL ends with the following words, freely adapted by the author of this paper from the writings of Sigmund Freud and Konrad Lorenz, that scroll on the screen:

Once the words were spells

And even today they have retained their magical power:

Words make you happy or push into despair;

With the words the teacher conveys his knowledge;

With the words the speaker drags the audience and determines their judgments and decisions;

With words you love or hate each other;

With words we construct equations that mock the words themselves;

With words we hypothesize the schemes and define the concepts and construct the image of reality.

But even today the realist looks at the external reality without realizing that he is the mirror,

While the idealist looks only in the mirror, turning his back on external reality;

So they do not see that the mirror has a reverse, a non-reflective face,

Almost a two-faced Janus, which places it on the same level as the elements it reflects.

Discussion and Conclusions

Physics presents us with representations of the world around us that, although changing in time, are in the end sufficiently coherent in themselves. Each theory, related to a correspondence principle, smoothly merges into a "larger" one that, thus, provides a new representation that we understand as deeper and more rigorous than the previous one.

It is worth pointing out that very word "representation" brings us naturally closer to the Theater. To represent means to present again; not so much to present the same thing several times, but to show with new eyes what is already present to an audience. It is in this role that one sees/discovers new things. From this point of view, science is like art. In fact, *"Art does not reproduce what we see, rather allows us to see"* said Paul Klee; which echoes Einstein's words: *"A great thought begins by seeing something in a different way, with a shift of the mind's eye"*. Therefore, theater, considered as the prototype location of all representations, can be, through the eye of the physicist, also the place where the image of physics (or rather, some of the images interiorised by physicist) is shown in its authenticity. And since, as the philosopher Henri Bergson says, *"The eye can see only what the mind is prepared to understand"*, a theatrical representation of physics designed for a non-specialist audience cannot be structured only like a "common" lesson (although we know that, in general, lessons have many theatrical aspects). In fact, the eyes of the physicists who make *"standard"*—even if *"well prepared and beautiful"*—lessons are too far away from the *"common"* eye in order to create direct empathy with the spectators.

Physicists are so deeply immersed in their discipline, in which they literally live, that they often think that it is enough to look, perhaps with just a touch of their guidance, at the things around us for everyone to able to see what they are able to see. However, us researchers in physics education know that it does not happen this way. People must learn to look, in depth, with different eyes and use new conceptual tools. These new tools are specifically designed experimental tasks and special conceptual structures. The language of physicists is a mathematical, abstract, sophisticated language. Its aesthetic canons are not those of the natural wonder, nor even those, though abstract, but certainly more common, of a painting or of a beautiful poem. Therefore, it takes patience and passion in order that people—students for example—if curious, if longing for knowing, can develop their understanding. In our experience, theater of physics and, more specifically, ALs can be very effective in achieving this goal [12].

We observed that ALs have been projected to be "more" than lessons. In fact, we believe that the potentialities of "TANTIT" are rooted in the way that it speaks to the audience through many channels. Not only the rational (rather obvious) channel mostly used during lessons, but also through passion, poetry, music and dance, and the enjoyment and magic that can born in an unconventional theatrical environment. ALs aim to be a way of helping students realise that, grounded on the formal education gained at school, we can find the search for simplicity and beauty; a beauty that is too often left in the hands of the artists alone or, worse, that often people find only in informal activities [8].

"TANTIT" has a meta-disciplinary goal: that is, of discussing the meaning of physics understanding and explanations. Therefore, it can be "used" whenever a teacher finds it worthwhile. For instance, it can be a part of a multidisciplinary activity connecting philosophy and physics and, maybe also art, while discussing the changing *weltanshauung* (worldview) in the history of mankind. Or it may be used in a meta-disciplinary discussion about the challenges of man facing nature and nature facing man … Or it can be used as an inquiry based activity with students to stimulate the use of theater in the classroom. In all cases, TANTIT is intended to be a bridge to connect the two shores of formal and informal physics education.

References

1. Carpineti, M., Cavallini, G., Giliberti, M., Ludwig, N., Mazza, C., Perini, L.: Let's throw light on matter: a physics show for primary school. Nuovo Cimento B **121**, 901–911 (2006). https://doi.org/10.1393/ncc/i2007-10022-7
2. Carpineti, M., Cavinato, M., Giliberti, M., Ludwig, N., Perini, L.: Theatre to motivate the study of physics. JCOM **10**(1), 1–10 (2011)
3. Barbieri, S., Carpineti, M., Giliberti, M., Rigon, E., Stellato, M., Tamborini, M.: "Good Vibrations": a workshop on oscillations and normal modes. Il Nuovo Cimento **38C**, 1–10 (2015). https://doi.org/10.1393/ncc/i2015-15086-1
4. Cavallini, G., Giliberti, M.: La lezione dalla Fisica Quantistica. Epistemologia **31**(2), 219–240 (2008)
5. Carpineti, M., Cavallini, G., Giliberti, M., Ludwig, N., Mazza, C.: Let's shine light on the matter: a physics show for primary school. In: Modelling in phisIcs and Physics Education: Proceedings of the GIREP Conference 2006, Amsterdam, Netherlands, pp. 819–822 (2008)
6. Carpineti, M., Giliberti, M., Ludwig, N.: Luce. In: Attori del sapere. Un progetto di teatro, scienza e scuola. Scienza Express edizioni, Milano, pp. 209–218. (2011). ISBN 978-88-96973-20-2
7. Carpineti, M., Giliberti, M., Ludwig, N.: Tracce. Fisici in teatro. In: Attori del sapere. Un progetto di teatro, scienza e scuola. Scienza Express edizioni, Milano, pp. 241–252 (2011). ISBN 978-88-96973-20-2
8. Giliberti, M.: Fisica a teatro. ARACNE editrice (2014)
9. Cavinato, M., Giliberti, M.: La Fisica in un Laboratorio, di Teatro. Scienzainrete, (2010). http://www.scienzainrete.it/contenuto/articolo/La-fisica-un-laboratorio-di-teatro. Last accessed: 2018/08/13
10. Carpineti, M., Ludwig, N.: "Fisica e Teatro: una scommessa vinta dal laboratorio SAT". Scienzainrete, (2010). http://www.scienzainrete.it/contenuto/articolo/Fisica-e-Teatro-una-scommessa-vinta-dal-Laboratorio-SAT. Last accessed: 2018/08/13
11. TEMI European Community's Seventh Framework Programme (FP7/2007–2013) under grant agreement Number 321403 (2012). http://teachingmysteries.eu. Last accessed: 2018/08/13
12. Carpineti, M., Giliberti, M., Ludwig, N.: Theatre as a means of supporting the teaching of science—Light Mystery—A script with added comments (2016). ISBN/EAN 978–94-91760-19-8
13. Carpineti, M., Giliberti, M.: Il teatro di fisica come primo passo verso l'Inquiry based Science Education nel progetto europeo TEMI (The theatre of physics as a first step towards Inquiry based Science Education in the European project TEMI). Giornale di Fisica **55**, 339–347 (2014). https://doi.org/10.1393/gdf/i2015-10208-9
14. Escobar, S.: L'infinito e la ricerca della semplicità, translated by the author (2015). https://www.alfabeta2.it/2015/02/28/linfinito-e-la-ricerca-della-semplicita/. Last accessed: 2018/08/13
15. Giliberti, M., Persano Adorno, D., Pizzolato, N., Battaglia, O., Fazio, C.: Augmented Lectures: Benefits of Supporting Physics Teaching with the Theatre. In: e-Proceedings of the GIREP-ICPE-EPEC Conference "Bridging Research and Practice in Physics Teaching and Learning" 3–7 July 2017 (2018)

Pedagogical Strategies to Increase Students' Engagement and Motivation

Claudia De Grandi⑩, Simon G. J. Mochrie⑩ and Rona Ramos⑩

Abstract With the goal of improving accessibility and inclusivity, we have implemented a series of pedagogical interventions to increase students' engagement and
learning in a large introductory physics course. Active learning and collaboration
were promoted during the lectures via several in-class strategies, including weekly
10-min quizzes, clicker questions, peer discussions, and group quizzes. Metacognition was encouraged by prompting students to reflect on their mistakes via optional
exam correction assignments. In addition, self-motivation was prompted by reflections on mindset (fixed versus growth) and the consequences of in-class multitasking.
The course also features two key outside-of-class resources, namely peer tutoring
and evening study-halls. Here, we discuss students' responses to these different
strategies and highlight the successes and challenges of increasing engagement and
performance. Among all of these components, study-halls emerged as the most successful tool to foster collaboration among students and peer learning, and to bring
faculty and students closer by permitting them to work together in an informal setting. Although our experience concerns an introductory physics course, geared to life
science and pre-medical students, our strategies are independent of subject and class
content, and therefore could be used in any university course, as well as adapted for
secondary schools.

Keywords Engagement · Group work · Metacognition · University physics

C. De Grandi (✉) · S. G. J. Mochrie · R. Ramos
Yale University, New Haven, CT 06511, USA
e-mail: degrandi@physics.utah.edu

S. G. J. Mochrie
e-mail: simon.mochrie@yale.edu

R. Ramos
e-mail: rona.ramos@yale.edu

C. De Grandi
University of Utah, Salt Lake City, UT 84112, USA

© Springer Nature Switzerland AG 2019
E. McLoughlin and P. van Kampen (eds.), *Concepts, Strategies
and Models to Enhance Physics Teaching and Learning*,
https://doi.org/10.1007/978-3-030-18137-6_19

Introduction

Introductory physics courses are generally a compulsory requirement for students seeking to pursue a career in life sciences or medicine. Among the major challenges in teaching such a course are (i) to make the material relevant to students' interests, (ii) to maintain students' motivation and engagement, and (iii) to create a welcoming and inclusive learning environment.

To tackle the first challenge, several ongoing efforts have been initiated in the past few years aimed at redesigning the traditional introductory curriculum to include biologically and medically relevant topics [1–3]. Within the context of a newly-designed course of this kind [4], here we describe the impact of several novel pedagogical interventions that go beyond the content of the course to address the psychology of learning and students' engagement and social belonging.

A great deal of evidence [5] supports the value of active learning and the fact that students learn better in a collaborative environment, where they can discuss, question and listen to their classmates' points of view. At the same time, social belonging—the sense of feeling welcomed by others and being able to positively relate to them—is fundamental to student well-being and productive cognitive processes [6], especially for minority students [7]. Finally, empowering students to be aware of their own learning process, that is, to possess metacognitive skills, is becoming increasingly viewed as important for students' learning success [8], especially in Science, Technology, Engineering and Mathematics (STEM) courses [9, 10].

With these factors in mind we have implemented several teaching strategies with the goal of building a collaborative and inclusive classroom environment and increasing students' motivation and metacognition. A number of these strategies have previously been widely promoted and adopted, including using of in-class clicker questions [11], adopting classroom spaces and seating arrangements that promote group work [12, 13], and implementing absolute grade lines [14]. We also employ peer tutors, similar to Learning Assistants [15] but without a required in-class presence, these are students who took the class the previous year and are hired to tutor current students, either one-on-one or in a group. Our grading rubric supports student improvement by dropping students' lowest problem set and in-class quiz scores, and rewarding strong performance in the final cumulative exam without unduly raising the stakes of this exam. We state these policies clearly at the beginning of the semester, so that students receive the message that success involves persistent hard work until they master the concepts.

In this paper, we focus on five teaching strategies, that also successfully promote collaboration, peer learning, and a sense of belonging, but that are more novel. We hope that other instructors may be interested in adapting these approaches to address the needs of their own students, and incorporate them into their own teaching practices. Specifically, we will discuss:

- Study-halls: creating an inclusive community of learners
- Group quizzes: learning through discussion with classmates
- Exam correction assignments: learning through careful dissection of mistakes

- Mindset: promoting motivation by reflecting on growth
- Encouraging focus by banning cellphone and laptop use in class.

We have implemented these strategies in a large (about 100–150 students) introductory physics course for life sciences majors at a research university in the United States. As described in detail by one of us [4], the course differs from a traditional introductory physics course because it includes topics relevant to premedical and biology students; for instance, after covering the traditional topics of kinematics, dynamics and energy, the course introduces basic probability, to be able to describe random walks and their application to diffusion in biology. Fluid dynamics is discussed in the context of the human circulatory system; mathematical modeling is introduced to describe the rate of change of quantities as a function of time, for instance the spread of a virus in a population, or the concentration of a drug in the blood. The student audience is composed mostly of pre-medical students (81%) and biological science majors (64%).[1] It is 67% female and ethnically diverse: 45% self-identify as white, 55% do not, including significant numbers from groups underrepresented in STEM (11% African American, 13% Hispanic).

Study-Halls: Creating an Inclusive Community of Learners

Our most successful intervention is the use of "Study Hall." Study halls are informal sessions, now offered three times per week, where students can come to work together on course assignments and get help and advice from the course instructors, teaching assistants, and peer tutors. Study-hall is held at a time and place that is convenient for students. For instance, for our mainly-on-campus students, study halls are held in the evenings at a location on campus easily accessible from their dormitory rooms. In a community college setting, one of us has implemented study halls immediately before or after class. The idea underpinning study hall is to create a space for students to work in groups with their classmates, and to foster an informal and relaxed collaborative environment between students and instructors. A venue with large tables where people can comfortably sit in variable-sized groups, and with whiteboards, is favorable. We staff each study hall session for a duration of 2 or 3 hours to provide the flexibility needed for students to be able to come at times convenient for their schedules, as well as to have enough time to work consistently on a significant part of their weekly homework assignment. Initially, study-hall was held only once a week, the night before the homework was due, and soon became a very hectic and busy night. Over time, we have eliminated traditional Teaching Assistant (TA)-led recitation sections in favor of increasing the number of study-halls to three per week. In this way, we encourage students to start working on their assignment early in the week, doing a part at a time, instead of scrambling to do the entire assignment the day before the assignment deadline. Early-in-the-week study halls have proven to be an incentive for many students to develop more organized study-habits.

---

[1] A student can be both pre-med and a biology major, the two categories are not exclusive.

Study-hall can also be viewed as a more inclusive way to hold office-hours. Instead of students having to come to the instructor's office, and face the pressure of a one-on-one meeting, now instructors come to a shared space, where students are comfortable to attend. Students can come individually or in groups; they can decide to talk to the instructors or can simply work in groups with their classmates. There is no agenda or requirement for study-hall, other than establishing a pre-defined time and place every week, where students are comfortable and can be sure to find help on course material. However, we recognize that study-hall goes beyond simply being a tutoring resource. It has proven a vehicle for building a learning community. Not only does it facilitate the formation of study-groups, numerous students also report that they made new friends at study-hall, whom they would have not met otherwise. The presence of the course instructors, instead of only teaching assistants or peer tutors, also contributes significantly to the success of study-hall. By making themselves available outside of class, the instructors communicate the value that they place on the work students do, and that they take personal interest in students' success. During study-hall, instructors sit at tables with students mingling with them. Naturally, most of the discussion is course focused, but, from time to time, we also discuss current events and other aspects of campus life. Thus, instructors learn about their students' interests, learn their names, and can even act as mentors on other aspects of their academic life. Students also come to know and trust their instructors. Initially, we noticed that a number of students—often students from underrepresented or disadvantaged groups—would come to study hall, but were nevertheless reluctant to approach the instructors or other course staff for help. To overcome this barrier, especially at the start of the semester, we are now proactive about getting to know and helping all students in the class. During the 2015–16 and 2016–17 academic years, 85% of students reported that they attended at least one or more study-halls per week, while 50% attended two or more per week.

To assess the impact of study-hall (and our other interventions), we gathered students' feedback through an anonymous survey administered at the end of the semester. In one question, students were asked:

Imagine you could write down a short paragraph to be given to next year's students in this course. Please write down what you would say to them to help them learn the most in the class.

The response rate was 89%. Answers were open ended, and they have been coded according to four categories:

(i) Students recommend attending study-hall (55% of responses).
(ii) Students recommend working with classmates and valued collaboration with peers (35% of responses).
(iii) Students explicitly advise on being comfortable asking questions and not being afraid to ask for help (31% of responses).
(iv) Students recommend talking and getting to know their professors (20% of responses).

Overall 40% of the responses addressed at least two of the four categories above. Here, we collect a few representative comments:

- ALWAYS go to study halls. Go to all of them if you can. They are honestly really helpful, especially when you have your instructor or the TAs helping you. It also gets you to collaborate with your fellow peers. You're all in this together, so why not work together? Don't be afraid to ask questions either, whether during study hall or in lecture. Without communication or attending study halls, you won't succeed in this class. Let the instructors help you!

- The best experience was the study halls. I made a bunch of new friends through it and I got work done I never could have figured out on my own. Go to the study halls!

- Interact with the people around you. Go to study halls and do the problem sets with your peers, but also don't be afraid to talk to the professors and get to know them as well. It makes a huge difference because once you're comfortable with the people around you, you're not afraid to ask questions during class. You're far less likely to feel lost or overwhelmed when you're comfortable enough to sense how you learn best and ask questions.

- [..] The fact that the faculty members are at the study halls really motivates me to attend. I am putting in a lot of work, but if the teaching staff are right there with me trying to help, I don't mind at all!

Group Quizzes: Learning Through Discussion with Classmates

One way to keep students up to speed with class material is to have regular low stake quizzes. In our course we have a 10-minute quiz at the beginning of every week, covering the material of the previous week. Students are given specific directions on how to prepare for each quiz. With adequate preparation the quiz will be straightforward for most students. However, if they do not study, a poor quiz score negatively impacts their overall grade. This weekly quiz incentivizes students to not fall behind with the material and also provides them with frequent immediate feedback on their understanding of new concepts. We administer the quiz on paper in class, and review the solutions immediately afterwards. In this way, we engage students in understanding when the questions are still fresh in their minds, offering an opportunity to clarify any missed concepts. From approximately ten quizzes per semester, the lowest three scores are dropped from the calculation of the final course grade. This policy ensures weekly quizzes are low stakes and that students are not overly penalized if they occasionally do not prepare well for a quiz.

About three times during the semester, a weekly assessment is given as a group quiz. Students are not informed ahead of time and they arrive in class prepared to take an individual quiz. Our version of group quizzes was inspired by the work of J. Ives and collaborators on group exams [16, 17] and adapted to our purposes. In a group quiz, students work in groups of three (two if necessary, but never more than three). They decide whom to work with before starting the quiz. Each student is given an individual white paper copy of the quiz and a single color paper copy is given to the group. As the quiz starts, they will spend a few minutes thinking on their own, then will discuss with their group. When the group comes to a consensus, they write their answers on the group quiz copy and submit it as a group. To allow time for discussion, group quizzes are allotted more time than an individual quiz—15 minutes instead of 10. When the group quiz is graded, each member of the group receives the same score. Students have the option of dissenting from the group and submitting an individual copy to be graded instead. Although this possibility is rarely used, we believe it is an essential option for students who may feel uncomfortable working in groups, and to provide a greater sense of overall control.

An advantage of group quizzes is the possibility to challenge students with questions that are more complex or require a higher level of cognitive skill than is reasonable for an individual quiz, or as a way for students to explore something they have not yet seen. Together with the use of in-class clicker questions and study-hall, group quizzes are one of the tools we use, especially at the beginning of the semester, to reinforce students' appreciation of the importance of collaboration. Because of the associated grade, students engage in vivid conversations with peers during group quizzes, and are eager to find the right answer. They come to appreciate the value of explaining things out loud, and the effectiveness of learning from each other. Interestingly, the average group quiz grade is usually higher than that of an individual quiz despite its more difficult questions.

Exam Correction Assignments: Learning from Mistakes

In most cases, in our collective experience, when graded assignments are returned, students look only at the grade, rarely reviewing the solutions or engaging in a thoughtful process of understanding what they did wrong. Our course has two midterm exams and one final exam. The final exam is cumulative and counts for 30% of the course grade, while the midterm exams each cover about one third of the course and count for 15% of the course grade. In order to promote student metacognition and bolster the role of formative assessments in learning, we have introduced an optional midterm exam corrections assignment [18]. This assignment has a dual goal: to encourage students to carefully review the provided solutions and reflect on their mistakes; to remove some of the pressure of the midterm exams by giving students that do not perform well an opportunity to improve their scores.

The procedure adopted for correcting exams is as follows. The midterm exam is graded and handed back to the students a few days after the exam was taken. Typed solutions to each part of the exam are also provided shortly after the exam. The exam is graded with clear indication where points were lost. From the day exams are returned, students have one week to submit an optional exam correction assignment. Depending on their performance and time commitment, they can decide to submit corrections for the entire exam or only for some problems or parts of the exam. If the assignment is fully completed students can gain back 30% of the points they lost on each part of the exam and for which they have submitted corrections.

Mistakes on the exam fall into three categories: conceptual, clerical or an omitted problem in which students left the question blank. In the exam correction assignment, students must first identify the category of their mistake. For conceptual mistakes, we follow closely the procedure introduced by Henderson and Harper [18]. First, students must identify what mistake was made and, more importantly, describe the erroneous reasoning that led to the mistake. Second, students must discuss the generalized concept or understanding learned in the dissection of the mistake that can be applied in other appropriate contexts beyond the specific problem. Clerical errors are mistakes due to lack of attention and not a misunderstanding of physical concepts. For this type of error, students describe the clerical error and how it led to an incorrect response on the exam. For omitted questions, students must review the solution and describe in words the logic and concepts behind the steps in the solution.

While working on this assignment, students can get help from instructors, teaching assistants and peer tutors at study-halls and office hours, they can work with classmates and consult any resource. We find that speaking to students one-on-one and listening to their thought process is one of the best ways to help them learn through this assignment, so an additional study-hall is offered during the week of the corrections. Exam corrections constitute a very insightful assignment both for the students as well as for the instructors. For many students, some concepts are only fully grasped while going through the corrections. For instructors, some misconceptions are only revealed by listening to students describe their thought processes.

Our exam correction assignments are atypical in a physics course, since they require considerable writing and metacognitive skills. They are very different from the usual assignments (such as quizzes and problem sets) that students are regularly required to do. Because of the metacognitive aspect, we recommend implementing exam corrections only after students have practiced on a lower stakes assignment. Therefore, early in the semester, we offer the possibility of submitting corrections for one of the weekly quizzes as a "warm-up".

Exam corrections require a significant time commitment from both students and the teaching staff. Grading narrative corrections can take a considerable amount of time. In order to streamline the grading, we assess corrections on a 0–2 scale, with 2 corresponding to full credit, 1 to half credit, and 0 to no credit. This way students can gain back either 30, 15% or none of the points they missed in a part of the exam. Furthermore, to limit the amount of writing to review, we set a maximum word limit for each part of the assignment. The word limit requires students to be concise and to the point, which, we believe, is also indicative of full understanding.

Given the incentive of boosting their exam grade, not surprisingly a large fraction (70–90%) of the class submits exam corrections. Over the four semesters during which we have implemented this assignment, the overall increment of the average midterm grade following corrections is 3–4%. We believe this assignment is a tool that removes some exam anxiety from the students and requires metacognitive skills, that we hope student will develop continuously over the course of the semester. We have gathered students' feedback about exam corrections. Here, we collect a few significant comments:

- I think this is a fantastic aspect of the course. It really makes you think deeply about the reasons you made mistakes rather than just correcting your solution to get points back. It made me re-evaluate my test taking strategies and led to a much better performance on midterm 2.

- I think that the chance to make exam corrections is one of the most helpful learning tools I've ever encountered at this school. As someone who often underperforms on tests due to the stress of a timed exam environment, I really appreciate the opportunity to make up for mistakes. Ironically, the knowledge that I could go back and correct an exam has actually taken some of the pressure off and allowed me to perform better during the exam itself. I've learned a great deal from the process of corrections. For instance, I'm able to see trends in the kind of conceptual errors I'm prone to make (many of them involve not reading the given information carefully enough to know how to flexibly apply a formula). My corrections on Midterm 1 (concerning how to graph the derivative of a function) allowed me to practice a concept that helped me perform better on Midterm 2, and without the correction process I probably wouldn't have done so.

Mindset: Promoting Motivation by Reflecting on Growth

Recently, it has been shown that having studied calculus in high school—likely a proxy for mathematics skills more generally—is the best single predictor of successful completion of a STEM degree [19]. Introductory physics courses, in particular, are often intimidating for non-physics majors because of physics' reliance on mathematics. Many of these students claim to be "not a math person," letting their previous mathematics experiences frame their expectations for physics. Unchallenged, such prejudices can limit students' chances of having a positive learning experience in physics. As instructors, therefore, it is our duty to try to mitigate students' preconceptions by facilitating a learning environment that promotes learning growth and rejects the idea of innate talent or skills.

We start the semester by challenging students' preconceptions of fixed mathematical ability or deficiency by discussing the theory of mindset developed by Dweck [20] and the learning and life outcomes resulting from a fixed mindset versus a growth mindset. These findings are succinctly and efficiently summarized in a 10-minute TED talk by Eduardo Briceno [21] that we either show in class the first day, or we ask the students to watch at home within the first week of class. To assess the impact of this intervention, we gathered students' feedback through an anonymous survey administered half-way in the semester. Students were asked:

On Lecture 1 of this class, we discussed the psychology of learning and the difference outcomes of having a growth mind-set versus a fixed mind-set and we watched together the Ted Talk "The Power of belief" by Eduardo Briceno. Now that we are in the middle of the semester what do you think is your current mind-set attitude towards this class?

The response rate was 87%. Answers were open ended, and they have been coded according to three categories:

(i) Students explicitly say they have a growth mindset or show a positive attitude towards the class. (61% of responses)
(ii) Students explicitly say they have a fixed mindset or show a negative attitude towards the class. (34% of responses)

(iii) Students are undecided, do not make any strong statement about their attitude or the class. (5% of responses)

Overall the majority of the students said they have a growth mindset. Some students found the video very inspiring and many said they were trying to transition to a growth mindset and to challenge themselves more in class. Here below we collect a few representative comments out of the 61% coded according to (i):

- I really enjoyed that video and I recommend it be shown at the start of the class every year. The video helped me do better and have a more positive mindset for all of my classes, not just this one. This is especially true since I did not consider myself a physics genius by any means but am currently exceeding my own expectations for the class.

- I definitely have a growth mind-set, although I think I came in with a fixed mind-set and was lying to myself out of shame. I did not perform well at the beginning of this semester, and I realized that the problem was myself. I was not giving this class the time that it deserved, nor was I giving myself the time that I needed. In realizing this, I am pushing myself to do better, to ask more questions, and to make efforts to understand more.

- My mindset feels like a growth-mindset. This course is really teaching me a new way to think!

In the second implementation of this mindset intervention, we gathered students' feedback through the anonymous survey administered at the end of the semester. Students were asked (similar to the question already quoted in the Study-hall section):

Imagine you could write down a short paragraph to be given to next year's students in this course. Please write down what you would say to them to help them have the best experience in the class.

The response rate was 88%. 35% of the responses explicitly mentions that struggling and being confused is a normal part of the learning process, and reflects an optimistic, resilient and growth mindset towards the class. Here below we collect a sample of these comments:

- To have the best experience in this course, you must first recognize that you are going to be confused most of the time. THAT IS OKAY. As the instructors say, confusion is the first step in learning. Work with your classmates, talk to your professors, go to office hours, work on psets in study halls, learn to ask for help and you will actually enjoy the class.

- Get a good, small group of friends and agree to work your hardest together. Help each other out and encourage each other. This class is a group effort, so embrace that aspect of it and do well together!

- Don't let a novel idea scare you. I think a lot of physics is incredibly counter intuitive. I think the people who are successful in the class are the people who are okay with being really confused for a while about the concepts but then struggling through them until they get to the heart of the problem.

We believe it is important to acknowledge the challenges that students face in our course, and to encourage them to not give up early on. Interventions of this kind may be crucial for the motivation and success of some students in our courses, while they will not hurt the others.

Encouraging Focus by Banning Cellphone and Laptop Use in Class

Access to electronic devices is increasingly recognized as a major distraction in the classroom [22]. The use of laptops has been shown to be detrimental for learning not only for the laptop user but also for nearby peers [23, 24]. In our course, we ban students from using all electronic devices for the duration of the lecture. To persuade students of the merit of this policy, we devote a part of the first class of the semester to (i) presenting evidence that devices hinder learning. Specifically, we show results from Sana et al. *Laptop multitasking hinders classroom learning for both users and nearby peers* [23] and encourage the students to take handwritten notes by arguing this is the first step towards building knowledge and learning [25]. More persuasively, we (ii) implement an activity in which students can see for themselves the deleterious effect of multitasking. This is a five-minute activity in which students are asked to write two sentences as fast as possible on a piece of paper. The first time they write the two sentences one after the other one, the second time they switch back and forth between the two sentences, interlacing the characters from each sentence [26]. Students will find that the second task takes about 1.5 times longer than the first. This makes it clear to students how switching back and forth between two different tasks—i.e. multitasking—takes more effort, concentration and time.

At the beginning of each class we always remind students to put away their phones and laptops and encourage them to stay focused for the remaining of the class. We have gathered students' feedback on this no-device policy via the anonymous end-semester survey. Students were asked in a multiple choice question to express their opinion on this policy. The answers collected were divided as follows: 49% were in favor, 40% were indifferent, 11% were against it. Below we collect a few examples of some optional follow up comments from students in favor of this policy:

- This should be a policy in every class.
- YES. I really really appreciated that […] Not only is it visually distracting but it also puts a niggling thought in my head that I am wasting my time by being engaged in the class […]
- I think that even the most well-meaning student gets bored/loses focus in class and when you have access to the internet, that's a recipe for disappearing into the rabbit hole.
- It's annoying sometimes when I see others on Facebook, and I hate it when I hear people typing in lecture. Also, physics is definitely better when you're writing things down.

Student Feedback on Class Atmosphere and Support

The strategies that we have described foster a collaborative and inclusive class community, in which students feel supported by their instructors and classmates. As a result, we see that students collaborate with each other instead of competing, that they are motivated to work hard, and that they perform well. To assess students' perspective on the class climate and community, we gathered feedback about these aspects as part of our anonymous end of semester survey. In many cases, students report that they make friends in the class, and that they enjoy the class, contrary to their expectations on both counts. A number of students recognize their own personal growth through the class. Generally, they are surprised to find such a welcoming atmosphere in a STEM course, in which they are, in most cases, enrolled solely to fulfill a compulsory requirement of their major or of admission into medical school. The data that follows is the average over three semesters of the course to the following questions:

- *How did you find the atmosphere of this course overall?* Multiple choice answer percentages: hostile 1%, not welcoming 3%, fine 30%, welcoming 61%, a love-in 5%.
- *Did you feel supported and cared about by your classmates?* Multiple choice answer percentages: Definitely yes 25%, Probably yes 47%, Might or might not 18%, Probably not 8%, Definitely not 2%.
- *Did you feel supported and cared about by your instructors?* Multiple choice answer percentages: Definitely yes 51%, Probably yes 35%, Might or might not 9%, Probably not 4%, Definitely not 1%.
- *How does the class community and climate that you experienced in this course compare with other STEM classes you have been taking at this university?* Multiple choice answer percentages: Much better 46%, Slightly better 30%, About the same 16%, Slightly worse 4%, Much worse 4%.
- *How does the class community and climate that you experienced in this course compare with other non-STEM classes you have been taking at this university?* Multiple choice answer percentages: Much better 21%, Slightly better 27%, About the same 36%, Slightly worse 10%, Much worse 6%.

In the following we collect a few representative comments from students' anonymous feedback on their perception of the class atmosphere:

- honestly! the best class I've taken so far in terms of community and friendliness and accessibility.
- I love that the class encourages you to collaborate with peers and work together rather than creating an environment in which we're "competing" with one another.
- I got to know the professors so much better than any other pre-med class. What I most dislike about STEM classes is there are a lot of introductory courses where you sit so far from the professor they need a microphone, but it really meant a lot that the professors learned everyone's names and seemed eager to interact with us.

Conclusion

We have presented a series of classroom interventions and teaching strategies aimed to increase students' engagement in the class, metacognition and a sense of belonging. We have gathered evidence that such strategies foster a welcoming community where students are able to make friends, enjoy working in groups, feel supported by the teaching staff and are encouraged to ask for help. As a consequence, students show a hard-working and motivated attitude towards course work, and perform at high levels in a difficult course, while also reporting an enjoyable experience in the class. Learning is not just about achieving the learning goals of a course. Just as essential is the environment in which learning takes place and the beliefs and attitudes of the learner. Lead by social and psychological components of learning, we created an environment in which we demand a high level of achievement from our students, but students still enjoy and embrace the challenge as a group, stay committed to the course and grow into scientists.

References

1. Meredith, D.C., Bolker, J.A.: Rounding off the cow: challenges and successes in an interdisciplinary physics course for life science students. Am. J. Phys. **80**, 913 (2012). https://doi.org/10.1119/1.4733357
2. Redish, E.F., et al.: NEXUS/physics: an interdisciplinary repurposing of physics for biologists. Am. J. Phys. **82**, 368 (2014). https://doi.org/10.1119/1.4870386
3. Crouch, C.H., Heller, K.: Introductory physics in biological context: an approach to improve introductory physics for life science students. Am. J. Phys. **82**, 378 (2014). https://doi.org/10.1119/1.4870079
4. Mochrie, S.G.J.: Vision and change in introductory physics for the life sciences. Am. J. Phys. **84**, 542 (2016). https://doi.org/10.1119/1.4947003
5. Freeman, S., et al.: Active learning increases student performance in science, engineering, and mathematics. Proc. Natl. Acad. Sci. U.S.A. **111**, 8410 (2014). https://doi.org/10.1073/pnas.1319030111
6. Baumeister, R.F., Leary, M.R.: The need to belong: desire for interpersonal attachments as a fundamental human motivation. Psychol. Bull. **117**, 497 (1995)
7. Walton, G.M., Geoffrey, L.C.: A question of belonging: race, social fit, and achievement. J. Pers. Soc. Psychol. **92**, 82 (2007). https://doi.org/10.1037/0022-3514.92.1.82
8. Bartimote-Aufflick, K., et al.: The study, evaluation, and improvement of university student self-efficacy. Stud. High. Educ. **41**(11), 1918 (2016). https://doi.org/10.1080/03075079.2014.999319
9. Tanner, K.D.: Promoting student metacognition. CBE Life Sci. Educ. **11**(2), 113 (2012). https://doi.org/10.1187/cbe.12-03-0033
10. McInerny, A. et al.: Promoting and assessing student metacognition in physics. In: Proceedings of the 2014 Physics Education Research Conference (PERC), pp. 179–182, (2014)
11. Mazur, E.: Peer instruction: a user's manual series in educational innovation. Prentice Hall, Upper Saddle River, NJ (1991)
12. Beichner, R. et al.: Student-Centered Activities for Large Enrollment Undergraduate Programs (SCALE-UP) project. PER-Based Reform in University Physics. American Association of Physics Teachers, College Park, MD (2006)
13. Dori, Y.J., Belcher, J.: How does technology-enabled active learning affect undergraduate students' understanding of electromagnetism concepts? J. Learn. Sci. **14**(2) (2005). https://doi.org/10.1207/s15327809jls1402_3
14. Grant, A.: Why We Should Stop Grading Students on a Curve. https://www.nytimes.com/2016/09/11/opinion/sunday/why-we-should-stop-grading-students-on-a-curve.html. Last accessed 2018/12/13

15. University of Colorado, LA website: https://www.laprogram.colorado.edu. Last accessed 2018/12/13
16. Ives, J.: Measuring the learning from two-stage collaborative group exams. In: Proceedings of the 2014 Physics Education Research Conference (PERC), pp. 123–126 (2014)
17. Ives, J. et al.: Examining student participation in two-phase collaborative exams through video analysis. In: Proceedings of the 2016 Physics Education Research Conference (PERC), pp. 172–175 (2016)
18. Henderson, C., Harper, K.A.: The Phys. Teach. **47**, 581 (2009). https://doi.org/10.1119/1.3264589
19. Bressoud, D.M. et al.: Insights and Recommendations from the MAA National Study of College Calculus. MAA Press (2015)
20. Dweck, C.S.: Mindset: The New Psychology of Success. Random House, New York (2005)
21. Briceno, E: The power of belief. https://ed.ted.com/on/CJeESw7Z. Last accessed 2018/12/13
22. McCoy, B.R. (2016) J. Media Educ. **7**(1), 5 (2016)
23. Sana, F., et al.: Laptop multitasking hinders classroom learning for both users and nearby peers. Comput. Educ. **62**, 24 (2013). https://doi.org/10.1016/j.compedu.2012.10.003
24. Hembrooke, H., Gay, G.: The laptop and the lecture: The effects of multitasking in learning environments. J. Comput. High. Educ. **15**(1), 46 (2003). https://doi.org/10.1007/BF02940852
25. Mueller, P.A. et al.: The pen is mightier than the keyboard. Advantages of longhand over laptop note taking. Psychol. Sci. **25**(6), 1159 (2014). https://doi.org/10.1177/0956797614524581
26. Cranshaw, D.: The Myth of Multitasking Exercise. https://davecrenshaw.com/multitasking-example/. Last accessed 2018/12/13

The Dust Catcher: Discovering the Educational Value of the Historical Scientific Heritage

Antonio Amoroso⑩**, Matteo Leone**⑩**, Daniela Marocchi**⑩
and Marta Rinaudo⑩

Abstract The purpose of this work is to present a physics education project, conducted at the University of Turin, whose goal is studying the educational significance of the history of physics and, in particular, of the history of scientific instruments. In this project, the survey of the scientific collection of the Museum of Physics of the University of Turin is followed by a census of the collections of physics instruments of historical-scientific interest preserved by a sample of schools in Piedmont. This study discusses the third part of the project which is devoted to exploring the expectations of in-service teachers on the usefulness of the history of physics for educational purposes and to designing and testing hands-on activities on electric phenomena inspired by the historical devices preserved by the Museum of Physics. In this paper we present the preliminary results of this latest part of the project. The aim of this project is to show that the dusty equipment in the old Physics Cabinets of Universities and schools, if appropriately studied and re-designed, could once again serve an educational function—that is they might provide us with a better insight into student's prior knowledge and at the same time promote a better understanding of physical sciences.

Keywords History of physics · Historical scientific heritage · Museum of physics

A. Amoroso · D. Marocchi · M. Rinaudo (✉)
Department of Physics, University of Turin, Turin, Italy
e-mail: marta.rinaudo@unito.it

A. Amoroso
e-mail: a.amoroso@unito.it

D. Marocchi
e-mail: daniela.marocchi@unito.it

M. Leone
Department of Philosophy and Educational Sciences, University of Turin, Turin, Italy
e-mail: matteo.leone@unito.it

A. Amoroso · M. Leone · M. Rinaudo
Museum of Physics, University of Turin, Turin, Italy

Introduction

This paper is part of a Ph.D. physics education project devoted to the study of the educational significance of the history of physics. More specifically, this project aims at developing the relationship between the Museum of Physics of the University of Turin and the old Physics Cabinets in secondary schools in the Turin and Cuneo provinces, in the north western region of Italy. The general objectives of the project are to: (i) survey the scientific collection of the Museum of Physics, including the instruments not yet catalogued; (ii) conduct a census of the collection of physics instrument of historical-scientific interest preserved by the schools in Piedmont, with a focus on the oldest public and private secondary schools that are likely to have preserved scientific collections, i.e. classical lyceum and technical institutes; (iii) catalogue teacher's expectations about a historical approach and (iv) design inquiry-based educational activities based on the development of the physics instruments displayed in the Museum and in the other scientific collections identified. The overall goal of this project is to provide a better insight into student's prior knowledge and promote a deeper understanding of physical sciences.

The specific activities are performed in the schools, in collaboration with the (about) 20 participating teachers in the project, i.e. approximately one teacher in each of the oldest classical lyceums in Turin and Cuneo provinces. The type of activity carried out is largely dependent on the current state of organization of the local Physics Cabinets. The experimental activities inspired by the local collection are designed and tested after the collection has been well surveyed and catalogued; on the contrary, when the collection has not yet reached an adequate level of organization, the activities focus on a census of the instruments and on historical researches on the physics behind the instruments. In both cases the students are expected to be active protagonists of the designed intervention.

In this paper, we present the preliminary results of this project in relation to the following research question: Is it possible to rediscover the educational value of a dusty collection of scientific instruments?

History of Physics as an Educational Tool

The collections preserved by University-based physics museums are usually made of instruments originally acquired for teaching *or* research. Yet, the sad state of affairs of most of these collections is that these instruments are by and large unused *neither* in research *nor* in teaching. This state of affairs is made even more sad by the growing awareness in the science education community of the science education functions of science museums as well as of the advantages of introducing history of science topics into the teaching of science [1, 2]. History and philosophy of science (hereafter HPS) might be a useful tool to help with identifying, and possibly overcoming, the mental representation that students have of physical science topics [3]. HPS and the wider domain of the history of material culture, as represented by the collections of old scientific instruments in schools and universities, may also prove to be useful at the meta-cognitive level. It was indeed argued that the collaboration between school and science museum, i.e. between formal and informal education, might promote achieving both cognitive and emotional student outcomes [4, 5]. One of the teaching formats that has been elaborated and evaluated by the science education researchers is just "conducting historical (thought) experiments or replicating actual laboratory procedures, tracing the development of scientific methods, concepts and theories" [6–10]. Furthermore, "historical approaches in science education offer substantial benefits in enabling people to develop scientific literacy and an understanding and appreciation both *in* science and *about* science" [11].

However, as it was recently emphasized [12], "*despite the positive educational effects of HPS, an apparent change in science teachers' attitudes towards it and the availability of HPS teaching resources, its occurrence in science classrooms is limited*" [13, 1]. While some science teachers do see history as a tool for fostering process skills and for illustrating the procedural aspects of real science, "*they seem to lack the professional knowledge, epistemological background and confidence to use HPS to support conceptual learning and to reflect on the contexts and nature of science*" [12, 14, 15]. In addressing the goal of the research question about the educational value of a dusty collection of scientific instruments, this project exploits the scientific collection of the former Physics Cabinet of the University of Turin, Italy [16, 17]. These instruments, now preserved by the Museum of Physics of the University, are over 1000 and have been partly catalogued. About 45% of them are exhibited in 23 showcases in the corridors of the old Institute of Physics and in 23 showcases in the Wataghin Hall, the old library of the Institute (see Fig. 1).

Fig. 1 The "Wataghin Hall" at the 1st floor of the old Institute of Physics

Most of the collection is devoted to instruments on electricity, magnetism and optics. This outcome is largely the result of the growing interest in 18th and 19th century physics toward the emerging fields of electricity and electromagnetism. This was especially true at the University of Turin, where the research and teaching activity of physicists like Father G. B. Beccaria and Abbè Nollet, in the 18th century, G. D. Botto, in the 19th century, and the skill of instrument makers like E. F. and C. Jest, in the 19th century as well, much contributed to the collection of the Physics Cabinet. The specifics of the collection of the former Physics Cabinet of the University of Turin make therefore the present museum an ideal place to carry on an historical–educational research focused on electricity topics.

Teacher's Expectations About the Historical Approach

In order to investigate the motivations that drive (or discourage) the decision to use an historical approach to introduce scientific themes and concepts, we have been administering a Likert scale questionnaire since Spring 2017 to a sample of (mainly) secondary school in-service teachers participating to training seminars in physics education organized by the University of Turin. The data collection is ongoing, however the preliminary results obtained from a sample of 78 in-service teachers are of sufficient interest to warrant their presentation in this paper.

On a 1–5 scale (where 1 corresponds to complete disagreement and 5 to perfect agreement), most of the teachers were in agreement (average 4.1) with the statement that "*it is helpful to bring a historical approach to normal disciplinary teaching to show the technological evolution of instrumentation*" (see Fig. 2).

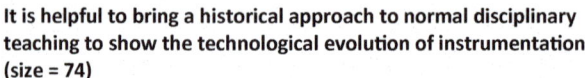

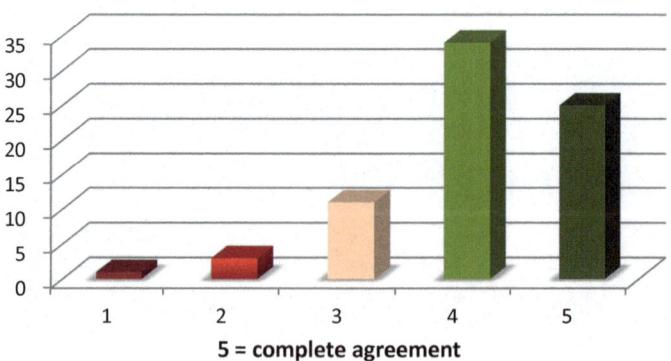

Fig. 2 Teacher's expectations about helpfulness of the historical approach

The questionnaire attempted, in particular, to address what are the possible limitations of the historical approach. Interestingly, on this issue most of the teachers showed disagreement with the statement *"it is not helpful to bring a historical approach to normal disciplinary teaching because I do not have the necessary preparation"* (average 2.2). According to this sample of teachers, a much more important factor is the lack of time: most teachers answered 2 (disagree) or 3 (indefinite) to the sentence *"it is not helpful to bring a historical approach to normal disciplinary teaching because it would take too much time"* (average 2.2). Finally, the questionnaire addressed the teacher's self-evaluation of their skills. Although most of this sample of teachers were confident that the preparation they had obtained by self-study research was adequate (average 3.6), most of them felt that they had not received an adequate preparation in history of physics for educational purposes during the undergraduate years (average 2.2).

On the whole, these preliminary results suggest that the teachers more active in the in-service training activities do not question the validity of the historical approach but, rather, fear that factors like the lack of time or the lack of adequate preparation in the undergraduate years might compromise the outcome of this approach [17].

Student's Prior Knowledge

With the goal of assessing the educational significance of history of physics and history of scientific instruments, we have been offering secondary schools (mostly grade-12 students from scientific lyceum classes), since Fall 2016, a one-day programme including a guided visit to the Museum of Physics and a workshop with easy inexpensive materials to make scientific instruments inspired to the collection of the Museum and, in particular, to the instruments about electricity.

Each session of work starts with a semi-structured questionnaire designed to detect student's prior knowledge about the physics contents later addressed in the workshop and, at the same time, to explore if history of physics can help us identifying aspects of such student's prior knowledge. Most items are indeed designed on the basis of the historical evolution of electricity studies with the goal of understanding if the possible difficulties experienced by students are historically rooted. In particular, the questionnaire explores student's prior knowledge about subjects like:

- the universal validity of the third Newton's law, i.e. also to the case of rubbing plastic rods (e.g. early fathers of electrostatics like G. Cardano and W. Gilbert, XVI century, believed that "amber is not attracted in turn by a straw") [18, 19];
- rubbing versus heating as the actual cause of the attraction in rubbing experiments (for Gilbert "amber does not attract by heat", however he did not discard the idea that the heat produced by friction is a relevant factor to obtain attraction) [19];
- the air as the agent responsible for the attraction/repulsion electrostatic effects (as it was argued by the Jesuit natural philosopher N. Cabeo) [20];

- the electrification as a phenomenon involving the space surrounding the rubbed object (e.g. Nollet) [21];
- the principle of charge conservation [22].

The above subjects are addressed by proposing to the student a number of experimental situations. By way of example, in one of these situations, a plastic rod is repeatedly rubbed with a wool cloth; after approaching the rod to some paper bits we notice that the bits stick to the rod. This problematic situation is followed by a number of statements about whom the student has to express his agreement or disagreement (true/false) and explain the reasons for his answer. Interestingly, 27% of the sample (size 54 students) answered that the attraction of the paper is caused by the heat produced by friction.

In another situation, a rubbed plastic rod is approached by a small ball of elder wood hanging on a cotton thread (i.e. a device like the many XVIII century electrostatics pendulums displayed in the Museum of Physics). We see the ball approaching the rod, touching it and then moving away. According to 18% of the sample, the ball approaches the rod because it is pushed by the air that tends to head towards the rod due to the effect that the rubbing has produced around the rod. Finally, 53% of the sample expressed disagreement with the idea that the ball moves away because on the rod and the ball there are electric charges of the same sign.

These very preliminary results suggest that suitably experimental situations, rooted in historically and much debated conceptual knots, could be very helpful in eliciting students' prior knowledge that otherwise might remain unexplored.

Hands-on Activities Inspired by the Museum of Physics Collection

The one-day programme offered to the schools includes a guided visit to the Museum of Physics where the original instruments are displayed, the administration of questionnaires to probe students' prior knowledge, and also includes a number of hands-on activities obtained by artifacts inspired on the historical instruments.

These activities are presently focused on electrostatics and electric current phenomena. Students experience what happens when one puts a rubbed plastic rod or rubber balloon close to a light object, what happens if other materials are rubbed, what happens if the rubbed object touches the light object, and so on. The students are therefore immersed directly in the motivating and complex phenomenology, typical of XVII and XVIII centuries, where electrification by rubbing, by conduction and by induction coexisted in an undifferentiated way. By learning how to build a Leyden jar, like those preserved by the Museum of Physics (see Fig. 3) with low cost, easy-to-find materials (see Fig. 4), the students deal with the phenomenon of electrification by rubbing and have an opportunity to take some "shocks" and appreciate the significance of the past attempts to "bottle" electricity with the precursor of the modern capacitor.

Fig. 3 Two sets of early 1800s dissectible Leyden jars preserved by the Museum of Physics of the University of Turin (inv. n.: 761)

Fig. 4 Construction of low-cost Leyden jars with plastic cups and aluminium foils

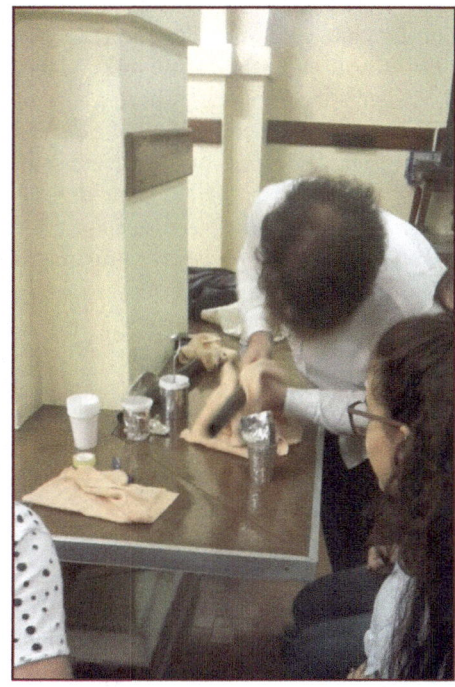

Fig. 5 Electroscopes with glass jars and their covers

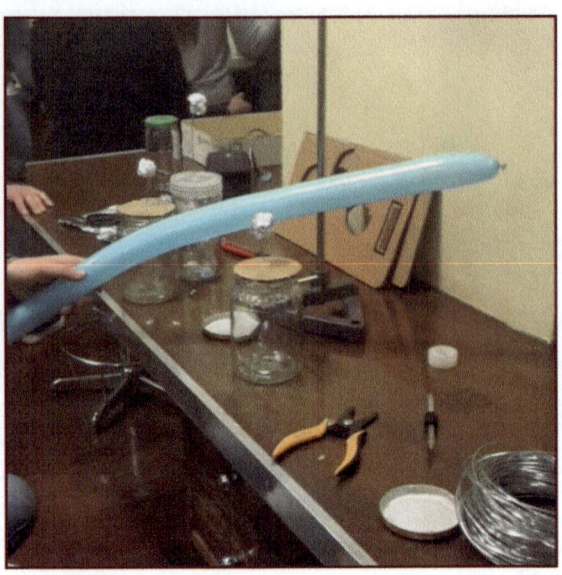

Through the construction of a rudimentary electroscope, inspired to those displayed in the collection, the students experience the struggle for arriving at a quantification of electricity and to the concept of electric charge (see Fig. 5).

Finally, through the voltaic pile and the continuous production of electric current the students experimentally observe how this apparatus differed from those developed in earlier times (an old version of a Zamboni dry cell is preserved in the Museum). The preliminary results of the student satisfaction questionnaire administered at the end of the one-day program show that students have the feeling of having "better understood the link between scientific discovery and evolution of the instruments" (83%). "*The desire to discover how scientific thought has evolved*" is increased as well (61%) (see Fig. 6).

Conclusions

In a very preliminary way, we may conclude that the activities of blowing away the dust out of the "dust catchers" of a typical Museum of Physics collection and developing artifacts inspired by these ancient devices is pretty motivating to students and well accepted by the teachers involved in this project. These activities, also, provide insights into the students' prior knowledge that might prove to be very useful in guiding the next activities in the formal education.

However, although the HPS approach is supported by a "*growing body of empirical evidence pointing out positive effects of science lessons enriched by historical information, experiments from past science and historically situated reflections on*

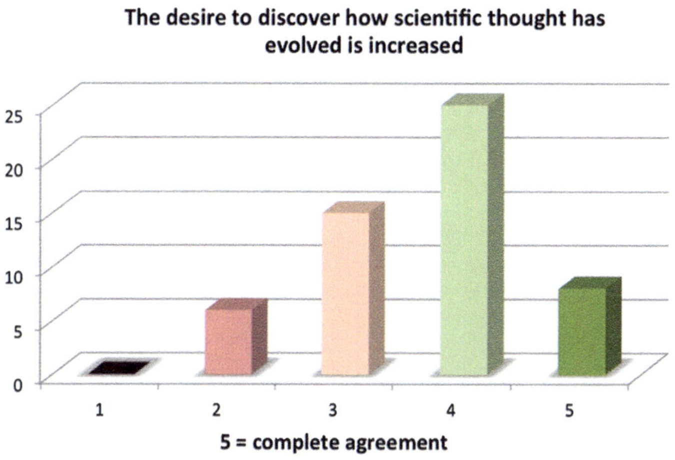

Fig. 6 Post-activity students' satisfaction after the one-day program at the Museum of Physics

the [nature of science]" [12], our survey about teacher's expectations also outlines a general lack of confidence of many science teachers in using the HPS approach. Many of the surveyed teachers believe indeed that they had not received an adequate preparation in history of physics for educational purposes during the undergraduate years.

In order to address this issue, the teachers participating to the project are currently engaged in training seminars about set up and organization of a school museum, analysis of students' prior knowledge and instructions on how to prepare a didactic activity using an historical approach. Of course, such an approach would benefit of a much more articulated learning path. The development of teaching pathways to deepen some aspects of the historical approach and to understand how to integrate them into curricular programs is currently ongoing.

References

1. Matthews, M.R.: Science Teaching: the Role of History and Philosophy of Science. Routledge, New York (1994)
2. Matthews, M.: Science Teaching. The Role of History and Philosophy of Science, 2nd edn. Routledge, London and New York (2015)
3. Leone, M.: History of physics as a tool to detect the conceptual difficulties experienced by students: the case of simple electric circuits in primary education. Sci. Educ. 23, 923–953 (2014). https://doi.org/10.1007/s11191-014-9676-z
4. Falomo Bernarduzzi, L., Albanesi, G., Bevilacqua, F.: Museum heroes all: the pavia approach to school-science museum interaction. Sci. Educ. 23, 762–780 (2014). https://doi.org/10.1007/s11191-012-9541-x
5. Filippoupoliti, A., Koliopoulos, D.: Informal and non-formal education: an outline of history of science in museums. Sci. Educ. 23, 781–791 (2014). https://doi.org/10.1007/s11191-014-9681-2

6. Binnie, A.: Using the history of electricity and magnetism to enhance teaching. Sci. Educ. **10**(4), 379–389 (2001). https://doi.org/10.1023/A:1011213519899
7. Heering, P.: The role of historical experiments in science teacher training: experiences and perspectives. Actes d'història de la ciència i la tècnica **2**(1), 389–399 (2009). https://doi.org/10.2436/20.2006.01.122
8. Höttecke, D.: How and what can we learn from replicating historical experiments? A case study. Sci. Educ. **9**(4), 343–362 (2000). https://doi.org/10.1023/A:1008621908029
9. Kipnis, N.: Theories as models in teaching physics. Sci. Educ. **7**(3), 245–260 (1998). https://doi.org/10.1023/A:1008697202578
10. Kubli, F.: Historical aspects in physics teaching: using Galileo's work in a new Swiss project. Sci. Educ. **8**(2), 137–150 (1999). https://doi.org/10.1023/A:1008613706212
11. Heering, P.: Science museums and science education. Isis **108**(2), 399–406 (2017). https://doi.org/10.1086/692689
12. Henke, A., Höttecke, D.: Physics teachers' challenges in using history and philosophy of science in teaching. Sci. Educ. **24**, 349–385 (2015). https://doi.org/10.1007/s11191-014-9737-3
13. Höttecke, D., Silva, C.: Why implementing history and philosophy in school science education is a challenge: an analysis of obstacles. Sci. Educ. **20**(3), 293–316 (2011). https://doi.org/10.1007/s11191-010-9285-4
14. Klopfer, L.E.: The teaching of science and the history of science. J. Res. Sci. Teach. **6**, 87–95 (1969). https://doi.org/10.1002/tea.3660060116
15. Wang, H.A., Marsh, D.D.: Science instruction with a humanistic twist: teachers' perception and practice in using the history of science in their classrooms. Sci. Educ. **11**(2), 169–189 (2002). https://doi.org/10.1023/A:1014455918130
16. Galante, D., Marino, C., Marzari Chiesa, A.: La collezione di strumenti di fisica dell'Università di Torino. Museologia scientifica memorie **2**, 287–289 (2008)
17. Rinaudo, M., Leone, M., Marocchi, D., Amoroso, A.: Il Museo: strumento di didattica della fisica? In: Bonino, R., et al. (eds.) Matematica e fisica nelle istituzioni (DIFIMA 2017). Graphot Editrice, Torino (2018)
18. Cardano, G.: De Subtilitate. Petreius, Nuremberg (1550). In: Forrester, J.M. (ed.) The De Subtilitate of Girolamo Cardano (English translation). ACMRS, Tempe (2013)
19. Gilbert, W.: De Magnete. Peter Short, London (1600). In: On the Loadstone and Magnetic Bodies (English translation). Wiley, New York (1893)
20. Cabeo, N.: Philosophia Magnetica. Franciscus Succhi, Ferrara (1629)
21. Nollet, J.A.: Recherches sur les causes particulières des phénomènes électriques. Guérin, Paris (1749)
22. Roller, D., Roller, D.H.D.: The Development of the Concept of Electric Charge. Harvard University Press, Cambridge (1954)